An Approach to Dark Matter Modelling

An Approach to Dark Matter Modelling

Tanushree Basak
Indus University, Ahmedabad, India

Morgan & Claypool Publishers

Rights & Permissions
To obtain permission to re-use copyrighted material from Morgan & Claypool Publishers, please contact info@morganclaypool.com.

ISBN   978-1-64327-132-3 (ebook)
ISBN   978-1-64327-129-3 (print)
ISBN   978-1-64327-130-9 (mobi)

DOI   10.1088/978-1-64327-132-3

Version: 20180801

IOP Concise Physics
ISSN 2053-2571 (online)
ISSN 2054-7307 (print)

A Morgan & Claypool publication as part of IOP Concise Physics
Published by Morgan & Claypool Publishers, 1210 Fifth Avenue, Suite 250, San Rafael, CA, 94901, USA

IOP Publishing, Temple Circus, Temple Way, Bristol BS1 6HG, UK

To my parents.

Contents

Preface

In the field of particle physics and astrophysics, one of the major unresolved problems is to understand the nature and properties of dark matter (DM), which constitutes almost 80% of the matter content of the Universe. This book will give a pedagogical introduction to the field of DM in general, and in particular to the model-building perspective. Starting from the evidence and need for DM, it goes into the deeper understanding of how to accommodate a DM candidate in a particle physics model. As a pre-requisite, an enthusiast reader must have elementary knowledge of quantum field theory, the Standard Model of particle physics and supersymmetry. Suggested books include An *Introduction to Quantum Field Theory* (1995) by Peskin and Schroeder, *Gauge Theory of Elementary Particle Physics* (1983) by Cheng and Li, and *Supersymmetry in Particle Physics* (2007) by Aitchison. The more advanced reader must be able to comprehend the calculation of relic abundances, annihilation cross-sections and scattering cross-sections.

This book focuses on teaching the basic tools for model building of DM, starting from the easiest and gradually moving through more difficult ones. Although there are plenty of DM models available in the literature, this book concentrates on the most important ones. The aim of this book is to motivate the reader to propose a new DM model complying with all observational constraints.

Tanushree Basak
Department of Science and Humanities
Indus University
Ahmedabad, India
May 6, 2018

Acknowledgments

My first and foremost acknowledgment goes to my editor Nicki Dennis, from Morgan & Claypool Publishers, for approaching me to write a book on this rather specialized topic. Prior to that I had never thought of putting down my concept and knowledge gained so far in a concise and pedagogical way. Special thanks goes also to Karen Donnison, who took care of the permissions for some of the figures, which saved me a lot of trouble.

A major part of this book is drawn from the work I have accomplished with my collaborators and my doctoral supervisor Professor Subhendra Mohanty. I am grateful to my supervisor for his invaluable guidance, encouragement and support throughout. I have immensely benefited from his insight and expertise in the subject. A very special thanks to my collaborators Ila Garg, Gaurav Tomar and Tanmoy Mondal for many fruitful and encouraging discussions which have helped greatly in shaping my thoughts.

I convey my gratitude to my parents for their constant support, encouragement and unconditional love. And last, but not the least, to my husband Koushik, who has always stood by my side with immense support and encouragement.

Author biography

Tanushree Basak

Tanushree received her Masters degree in physics from the University of Calcutta, India. As a graduate student, she was deeply motivated by the fascinating world of astro-particle physics and decided to continue her research in this field. She pursued her research interest by moving to the Physical Research Laboratory, Ahmedabad. She later earned her doctorate working on the phenomenology of particle dark matter models. She is currently working as an Assistant Professor in Physics in the Department of Science and Humanities at Indus University, Ahmedabad, India. Her primary research fields are supersymmetry, physics beyond the Standard Model, and dark matter and neutrino physics. Dr Basak has published several research papers and conference proceedings articles in peer reviewed international journals.

Abbreviations

BR	branching ratio
BSM	beyond the SM
CDM	cold DM
DM	dark matter
EW	electroweak
EWSB	electroweak symmetry breaking
FIMP	Feebly interacting massive particle
GC	Galactic Center
LHC	Large Hadron Collider
MSSM	Minimal Supersymmetric SM
NMSSM	Next-to-Minimal Supersymmetric SM
RH	right-handed
SD	spin-dependent
SI	spin-independent
SM	Standard Model
SUSY	supersymmetry
VEV	vacuum expectation value
WIMP	weakly interacting massive particle

An Approach to Dark Matter Modelling

Tanushree Basak

Chapter 1

Introduction

1.1 Why dark matter?

After the discovery of the Standard Model (SM)-like Higgs boson, the last missing particle of the SM, at the LHC [1, 2], a new era of particle physics has started. Although the SM provides very successful and precise descriptions of all experiments in particle physics, it has some theoretical shortcomings. The SM not only suffers from the hierarchy problem, but there remain many open questions such as: the mass of neutrinos; the origin of matter–antimatter asymmetry; what the invisible or 'dark' matter (DM) in the Universe is (of which there is five times the amount of all visible matter) etc. Here lies the pressing need to look beyond the SM in order to find an answer to some of these queries. The problem of DM is surely one of the most exciting and challenging open questions in physics.

The earliest identification of DM came from the velocity dispersions of galaxies within clusters. In 1933, Fritz Zwicky deduced the existence of a non-luminous constituent of the Coma cluster by observing the dynamics of the galaxies contained therein [3], famously conferring upon it the name of 'dark matter'. The only way the observed velocities of the cluster members could be reconciled with the virial theorem was to postulate that the cluster also contained another large, but unseen, mass component—dark matter.

Studies of cosmic-microwave-background (CMB) radiation play a crucial role in determining the DM abundance in the Universe. The CMB radiation observed today originates from the decoupling and recombination epoch. Temperature anisotropies (small inhomogeneities in the distribution) of CMB correspond to fluctuations of the matter density in the early Universe, which subsequently gave rise to the observed large structures. The power spectrum of temperature anisotropies (in terms of spherical harmonics) depends on cosmological parameters, which can be obtained by fitting the resulting spectrum with some underlying assumption of the cosmological model. Cosmological observations of CMB anisotropies, by the Wilkinson Microwave Anisotropy Probe (WMAP9) [4] and PLANCK satellite [5],

doi:10.1088/978-1-64327-132-3ch1

constrain the DM density (in units of the critical density) of the Universe to be $\Omega_{CDM}h^2 = 0.1149 \pm 0.0019$ (CDM—cold dark matter). From observations we know that our Universe consists of 71.4% dark energy, 4.6% luminous matter and 24% DM. Hence, the non-baryonic DM constitutes the majority—roughly 80–85%—of the matter in the Universe. The amount of matter and dark energy in the Universe can also be derived by analyzing Baryon acoustic oscillation (BAO) [6], supernova type Ia data [7, 8] or from the Lyman-α forests (neutral hydrogen clouds seen in absorption in quasar spectra) [9, 10].

Although the evidence favouring the existence of DM is extremely compelling [11], its nature mostly remains unknown. In this book, we focus on the extensions of the SM and propose various particle physics models to explain the experimental consequences of DM. In the pursuit of understanding DM phenomenology, many ideas from particle physics have been used, and in equal measure experimental observations from astrophysics and cosmology have been used to constrain ideas in particle physics. This interplay between theoretical particle physics and astrophysical observations lies at the core of this book.

1.2 Evidence for dark matter

1.2.1 Rotation curves

The earliest and as yet the most convincing evidence in support of the existence of DM came from the rotation curves [13, 14] of galaxies (a graph of circular velocities of luminous gas in a galaxy as a function of their distance from the galactic center). According to Newtonian dynamics the radial velocity of a galaxy in a cluster is given by

$$v(r) = \sqrt{\frac{GM(r)}{r}} \tag{1.1}$$

where, $M(r) = \int 4\pi\rho(r_1)r_1^2 dr_1$ and $\rho(r_1)$ is the density profile. The above equation shows that velocity should fall off as r increases, but instead the rotation curve in figure 1.1 shows a considerably flatter behavior, which suggests the existence of a halo with $M(r) \propto r$. Observations suggest that the velocities of distant stars in the galaxy remain roughly constant over a wide range of distances from the center of the galaxy, in contradiction with expectations based on the distribution of visible matter in the galaxy. Therefore, these rotation curves further demonstrated that the visible mass in those galaxies could not account for the observed circular velocities and hence postulated the presence of a large unseen component of mass inside the galaxy.

1.2.2 Gravitational lensing

According to general relativity, the presence of mass causes the space in its vicinity to curve. Clusters, galaxies and stars are massive enough to bend and focus light rays passing through their gravitational potential, which acts as a lens. As a result light from distant galaxies, quasars and stars are 'gravitationally lensed' [15] by other

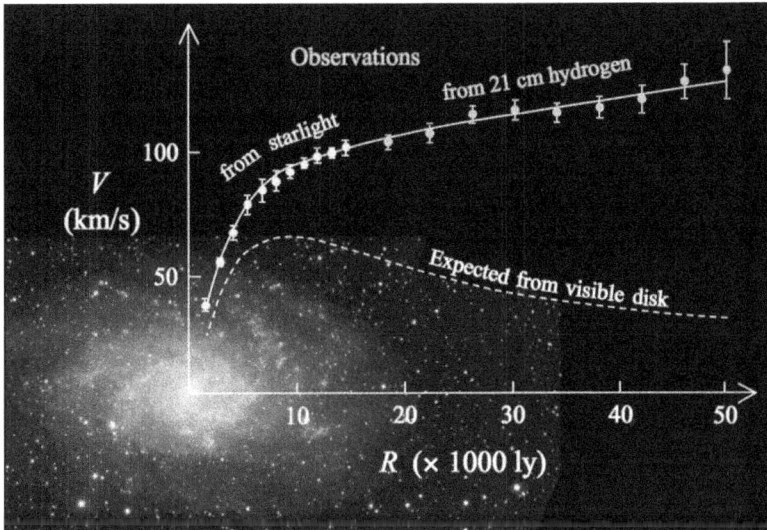

Figure 1.1. Rotation curve of dwarf spiral galaxy M33. The solid curve shows the observed velocity and the dashed curve shows the estimated contribution from the luminous disk. Reproduced with permission from [12]. Copyright 2000 IOP Publishing.

clusters and galaxies which lie in their path. The amount of 'lensing' depends on the mass of the object (the lens) causing this effect. It can therefore be used to determine the mass of astrophysical objects ranging from planets to galaxies and galaxy clusters.

A different example of gravitational lensing in the context of DM is that of the so-called bullet cluster [17], shown in figure 1.2. This figure shows the collision between a smaller cluster (bullet) with a primary cluster. The distribution of mass from the weak gravitational lensing (blue region) and that from the x-ray map (pink region), which consists of mostly baryonic matter, shows that the majority of matter in the two galaxies is non-baryonic and it is primarily in the halo region of the galaxy. Moreover, the separation in the two regions shows that while the gas clouds in the two galaxies exerted friction on each other resulting in the bullet shape of the rightmost cluster, the DM halos of the two galaxies pass through each other and the gas clouds without undergoing any collision. This strongly indicates the presence of a collisionless, non-baryonic DM halo.

1.2.3 Cosmological evidence

The Universe displays a very complex structure on a large scale. Galaxies of stars are part of a cluster, clusters of such galaxies are again part of superclusters which are then arranged into large-scale sheets, filaments and voids. Presumably, the pattern of galactic superstructure reflects the history of gravitational clustering of matter since the Big Bang. If DM were present during structure formation, it should have influenced the pattern of large-scale structure we see today. Large-scale structure surveys such as the 2DFGRS [18] and the SDSS [19] can also provide information

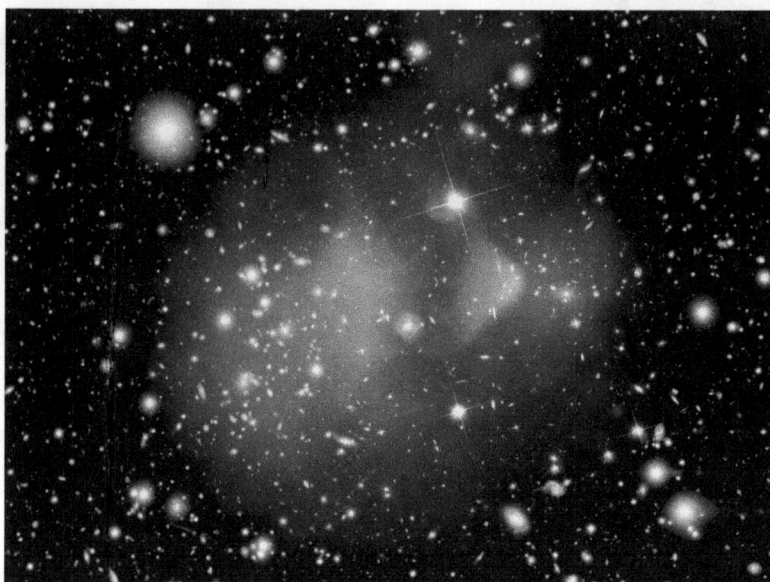

Figure 1.2. Composite image of the matter inside galaxy cluster 1E 0657-56, also known as the 'bullet cluster'. The blue region shows the lensing map while the pink region shows the x-ray data associated with the gas clouds. The clear separation between the x-ray and the lensing maps shows that most of the matter inside the galaxies is collisionless DM. X-ray data: NASA/CXC/CfA/Magellan/U.Arizona [16]. Optical data: NASA/ STScI [17]. Lensing maps: NASA/STScI; ESO/WFI/Magellan/U.Arizona [17]. Reproduced with permission from [17]. Copyright 2006 American Astronomical Society.

on the total matter density in the Universe [6, 20]. Large-scale cosmological 'N-body' simulations [21–24] demonstrate that the observed large-scale structure of luminous matter could only have been formed in the presence of a substantial amount of DM. A recent analysis in [6] indicates a total matter density of $\Omega_M = 0.29$. The Big Bang nucleosynthesis (BBN) data give the baryonic density as $\Omega_B = 0.04$. Combining this with the BBN result gives the non-baryonic DM density to be $\Omega_{CDM} = 0.25$.

References

[1] Aad G *et al* 2012 Observation of a new particle in the search for the Standard Model Higgs boson with the ATLAS detector at the LHC *Phys. Lett.* **B716** 1–29

[2] Chatrchyan S *et al* 2012 Observation of a new boson at a mass of 125 GeV with the CMS experiment at the LHC *Phys. Lett.* **B716** 30–61

[3] Zwicky F 1933 Die Rotverschiebung von extragalaktischen Nebeln *Helv. Phys. Acta* **6** 110–27

[4] Hinshaw G *et al* 2013 Nine-year Wilkinson microwave anisotropy probe (WMAP) observations: cosmological parameter results *Astrophys. J. Suppl.* **208** 19

[5] Ade P A R *et al* 2014 Planck 2013 results. XVI. Cosmological parameters *Astron. Astrophys.* **571** A16

[6] Percival W J *et al* 2010 Baryon acoustic oscillations in the Sloan digital sky survey data release 7 Galaxy sample *Mon. Not. Roy. Astron. Soc.* **401** 2148–68

[7] Perlmutter S *et al* 1999 Measurements of omega and lambda from 42 high redshift supernovae *Astrophys. J.* **517** 565–86

[8] Riess A G *et al* 2004 Type Ia supernova discoveries at $z > 1$ from the Hubble Space Telescope: evidence for past deceleration and constraints on dark energy evolution *Astrophys. J.* **607** 665–87

[9] McDonald P *et al* 2005 The linear theory power spectrum from the Lyman-α forest in the Sloan Digital Sky Survey *Astrophys. J.* **635** 761–83

[10] McDonald P *et al* 2006 The Lyman-α forest power spectrum from the Sloan Digital Sky Survey *Astrophys. J. Suppl.* **163** 80–109

[11] Trimble V 1987 Existence and nature of dark matter in the Universe *Ann. Rev. Astron. Astrophys.* **25** 425–72

[12] Bergstrom L 2000 Nonbaryonic dark matter: observational evidence and detection methods *Rep. Prog. Phys.* **63** 793

[13] Rubin V C and Kent Ford W Jr 1970 Rotation of the Andromeda Nebula from a spectroscopic survey of emission regions *Astrophys. J.* **159** 379–403

[14] Sofue Y and Rubin V 2001 Rotation curves of spiral galaxies *Ann. Rev. Astron. Astrophys.* **39** 137–74

[15] Bartelmann M 2010 Gravitational lensing *Class. Quant. Grav.* **27** 233001

[16] Markevitch M 2006 Chandra observation of the most interesting cluster in the universe *ESA Spec. Publ.* **604** 723

[17] Clowe D *et al* 2006 A direct empirical proof of the existence of dark matter *Astrophys. J.* **648** L109–13

[18] Colless M *et al* 2001 The 2dF galaxy redshift survey: spectra and redshifts *Mon. Not. Roy. Astron. Soc.* **328** 1039

[19] Tegmark M *et al* 2004 The 3-D power spectrum of galaxies from the SDSS *Astrophys. J.* **606** 702–40

[20] Cole S *et al* 2005 The 2dF galaxy redshift survey: power-spectrum analysis of the final dataset and cosmological implications *Mon. Not. Roy. Astron. Soc.* **362** 505–34

[21] Navarro J F, Frenk C S and White S D M 1996 The structure of cold dark matter halos *Astrophys. J.* **462** 563–75

[22] Springel V *et al* 2005 Simulating the joint evolution of quasars, galaxies and their large-scale distribution *Nature* **435** 629–36

[23] Diemand J, Kuhlen M and Madau P 2007 Dark matter substructure and gamma-ray annihilation in the Milky Way halo *Astrophys. J.* **657** 262–70

[24] Springel V *et al* 2008 The Aquarius Project: the subhalos of galactic halos *Mon. Not. Roy. Astron. Soc.* **391** 1685–711

Chapter 2

Particle dark matter candidates: constraints from observations

Despite the mounting evidence for particle dark matter (DM) in galaxies, clusters of galaxies and the Universe at a large scale, very little is known about the properties of the DM particle. The concept of DM does not find an explanation in the framework of the Standard Model (SM). The identity of DM is a question of utmost importance in both astrophysics and particle physics. Hence, plenty of extensions of the SM model were put forward with a motivation to introduce a suitable DM candidate. In this chapter, we discuss a plethora of DM matter candidates and constraints on them from direct and indirect detection experiments.

2.1 Dark matter particle candidates

According to observational evidence, we may conclude that DM particles are electrically neutral and interact with visible matter only weakly. Furthermore, to be in agreement with CMB data, most of the DM should be non-baryonic in nature. Therefore, potentially the only indication compatible with all the cosmological measurements is that DM is composed of non-baryonic, neutral and weakly interacting particles.

2.1.1 Weakly interacting massive particles

Among the non-baryonic candidates the weakly interacting massive particles (WIMPs) are the most favored and widely studied DM candidate, as they satisfy the astrophysical and cosmological criteria, and offer the possibility of detectable experimental signals. Examples of WIMPs include the lightest neutralino in supersymmetry (discussed in chapter 5), the lightest Kaluza–Klein (KK) particle [1, 2], scalar DM, an additional inert Higgs boson (discussed in chapter 3) [3–6] and right-handed (RH)-neutrino DM (discussed in chapter 4), etc.

doi:10.1088/978-1-64327-132-3ch2

2.1.2 Non-thermal relics

In the literature, several non-WIMP candidates (for reviews see [7–9]) have also been proposed, among which the most relevant are: sterile neutrinos, axions, gravitinos, axinos, DM from the little Higgs model, and superheavy DM particles or Wimpzillas.

2.2 Dark matter searches

2.2.1 Direct detection

In direct experiments one looks for these WIMPs passing through a detector and scattering off some nucleus (for reviews see, [11–13]). A variety of detectors [10, 14–22] designed to be sensitive to the nuclear recoils induced by collisions with WIMPs are currently collecting data, and have placed bounds on the WIMP-nucleon cross-section–WIMP-mass parameter space. This scattering can be detected and, if found, would be evidence for WIMPs in the Galactic halo. These studies also depend on astrophysical input such as, in particular, the local phase-space distribution of DM particles.

Scatterings of DM particles off nuclei can be detected using three main techniques: (1) scintillation, (2) ionization and (3) heat (phonons in crystal detector). Sometimes one or a combination of two such techniques are employed in a

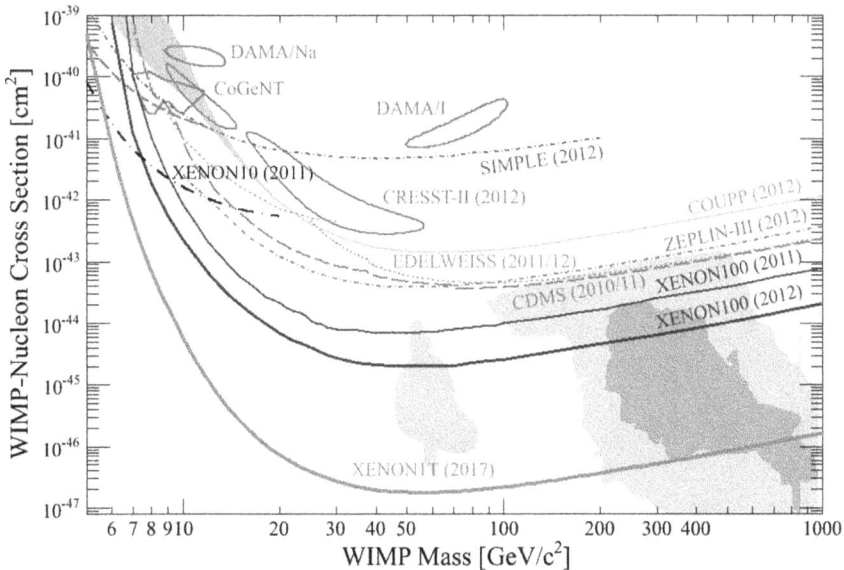

Figure 2.1. Spin-independent elastic WIMP-nucleon cross-section (σ_p^{SI}) as a function of WIMP mass (m_χ). The thick blue line shows the 90% CL for the latest XENON100 data. The limits from CDMS (dashed orange line), XENON(2010) (thin black line) and EDELWEISS (dotted pink line) are also shown. Finally, the 90% CL regions for CoGeNT (green) and DAMA (red, without channeling) are shown. Reproduced with permission from [10]. Copyright 2012 Springer.

particular experiment. Depending on the choice of signal detection technique, a variety of target materials can be employed in these direct detection searches.

A well-known example, the DAMA/LIBRA experiment [23], employs a scintillation detector technique and has been reported to observe an annually modulated DM-like signal for the last two decades. The current significance is at the level of 9.3σ [16]. The estimated mass of the DM particles ranges between 10 to 15 GeV or between 60 to 100 GeV depending on the actual nucleus involved in the scattering process (sodium or iodine, respectively) with a scattering cross-section of $\sim 10^{-42}$ cm^2. However, the DM interpretation of these results is in strong tension with null results published by some other collaborations: the first XENON1T limits [10, 24], the final LUX [15, 25] and the PandaX-II [26] limits, which excluded the annual modulation of DM interpretation of the effect claimed by DAMA/LIBRA. In the low mass region, there are limits from the CDMSlite [27] and XMASS [28] experiments (figure 2.1).

The most stringent current limit on spin-independent scattering cross-section σ_p for large DM mass comes from null results of DM searches from XENON1T [10, 24], which improved the previous limits of the XENON100 collaboration [14]. Also, the final LUX result [25] and that of PandaX-II [26] give important bounds on the cross-section.

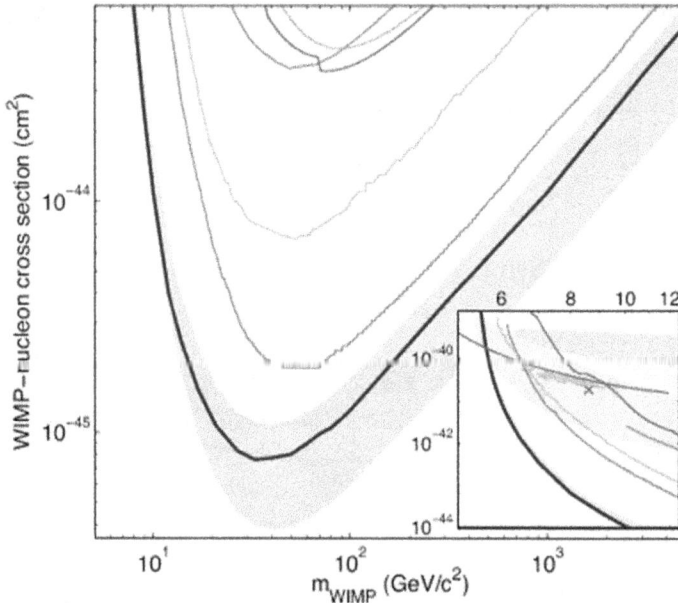

Figure 2.2. The LUX 90% confidence limit on the spin-independent elastic WIMP-nucleon cross-section (blue), together with the $\pm 1\sigma$ variation from repeated trials, where trials fluctuating below the expected number of events for zero BG are forced to 2.3 (blue shaded area). We also show Edelweiss II [22] (dark yellow line), CDMS II [20] (green line), XENON100 100 live-day (orange line), and 225 live-day [14] (red line) results. The inset (in the same axis units) also shows the regions measured from annual modulation in CoGeNT [18] (light red, shaded) and the DAMA/LIBRA allowed region [16]. Reproduced with permission from [15]. Copyright 2014 the American Physical Society.

2.2.2 Indirect detection

In indirect searches, one looks not for the WIMPs directly, but for signals coming from annihilation of two WIMPs. For example, their annihilations in the halo will result in a γ- and anti-proton-flux which can be searched for. A wide array of cosmic-ray and gamma-ray observatories—both in space and on the ground—are currently searching for indirect signals.

Cosmic rays play an important role in the indirect detection of DM. The charged cosmic rays, in particular positrons, anti-protons and anti-deuteron, are particularly promising tools for indirect searches due to their relatively low astrophysical background. There are many hints of DM annihilation in the high energy cosmic-ray spectrum of positrons, anti-protons and γ-rays. In 2009, the PAMELA Collaboration reported an excess in the positron spectrum [29, 30] which was subsequently confirmed by the Fermi-LAT [31] and the AMS-02 [32] experiments. DM annihilation may account for the excess positron flux seen in PAMELA [29, 30] and AMS-02 [32] experiments, shown in figure 2.3. Observation of γ-rays in Fermi-LAT [31] and Hess [33] give possible signals of DM annihilation into γ-rays.

Whether searching for annihilations or decays, the most promising targets are those with large DM densities and/or low astrophysical backgrounds. The Galactic Center (GC) would seem to be the most obvious target given its distance and DM concentration, but it is also one of the most difficult areas to work with because of its complex and poorly understood background. A few years ago, a tantalizing hint of DM was found in the analysis of the Fermi-LAT gamma-ray data [31] which revealed the existence of a peak at around 130 GeV coming from the vicinity of the GC. Later it was realized that this excess arose from statistical fluctuations, as its

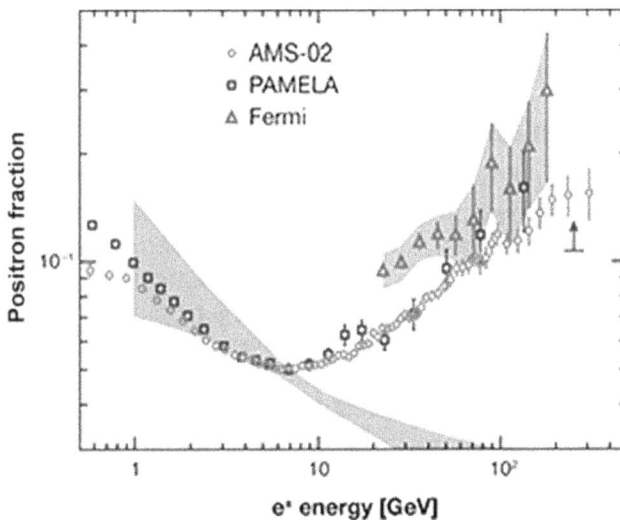

Figure 2.3. The positron fraction compared with the most recent measurements from AMS-02 [32], PAMELA [29, 30] and Fermi-LAT [31]. Reproduced with permission from [32]. Copyright 2013 the American Physical Society.

significance has diminished in subsequent improved analyses performed by the Fermi-LAT Collaboration.

In recent years, a possible excess in the x-ray emission near 3.5 keV was reported after analyzing the XMM-Newton data from observations of the Andromeda galaxy and various galaxy clusters [34, 35]. These observations were subsequently confirmed by other experiments such as Suzaku [36], Chandra [37], etc. There is a possible interpretation of the signal from the decay of sterile neutrino DM into photons.

2.2.3 Collider search on dark matter

Another possibility of searching for DM is through the production of DM in collider experiments from the annihilation of SM particles at sufficiently high energies. Any WIMPs produced at colliders will escape from the detector unnoticed. The most obvious collider WIMP signature is expected to be missing transverse energy (missing E_T), which refers to an apparent missing component of the total final-state momentum in the direction transverse to a collider beam. Monojet searches typically give the strongest constraints [39]. However, one can also hope to see DM in other channels such as monophoton searches, mono-W searches and mono-Z searches.

In the framework of Higgs portal DM, bounds from invisible Higgs decays can be translated into bounds on the WIMP scattering cross-section [44]. The resulting bounds apply only for $m_\chi < m_h/2$ and depend on whether the WIMP is a scalar, fermion or vector [45]. For effective operators inducing spin-independent interactions, LHC searches [38] are typically inferior to direct detection (except for very light masses), since the latter benefit from coherent enhancement (shown in figure 2.4). However, for spin-dependent interactions, direct detection cross-sections are not enhanced and LHC searches typically give the strongest bounds, as shown in figure 2.5.

2.3 Constraints on dark matter candidates

A good DM particle candidate should fulfill a series of important properties [8, 9, 46, 47] in order to provide a convincing explanation for all the observed phenomenology:

- It should be weakly interacting to ordinary matter, i.e. SM particles, and electrically neutral, i.e. with neither electromagnetic nor strong interactions.
- It should be long-lived with a lifetime larger than H_0^{-1} so that it will have survived since the time in the early Universe when it was created.
- It has to be cold, i.e. it should be non-relativistic when it decouples from the radiation in order to not erase the density fluctuations at galaxy scales. At most, the DM may be warm, with free-streaming lengths on the order of cluster scales of a few Mpc.
- It must be massive enough to account for the measured Ω_{DM}.
- It must be consistent with observations (BBN, relic density) and present constraints (direct and indirect detections).

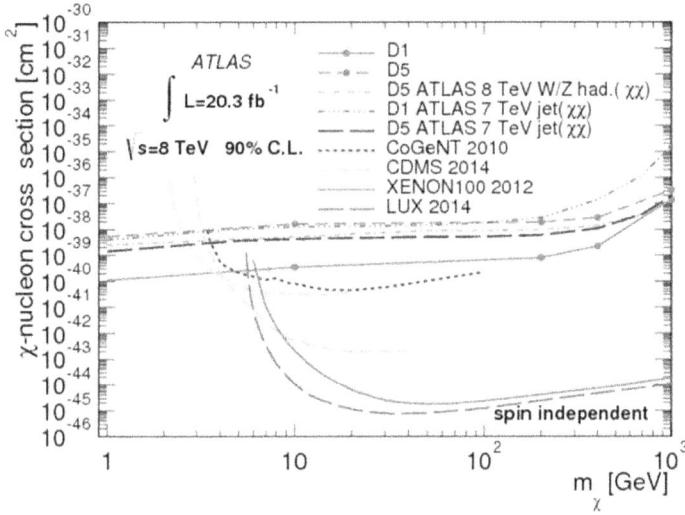

Figure 2.4. Observed 90% CL upper limits on the χ-nucleon scattering cross-section as a function of m_χ for spin-independent effective operators mediating the interaction of the DM particles with the $q\bar{q}$ initial state. The limits are compared with results from the published ATLAS hadronically decaying W/Z and $j + \chi\chi$ searches, CoGeNT [17], XENON100 [14], CDMS [19, 21], and LUX [15]. Reproduced with permission from [38]. Copyright 2014 the American Physical Society.

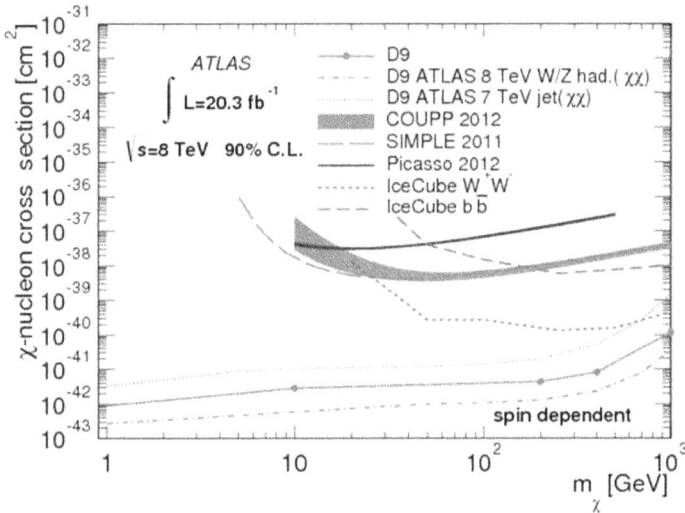

Figure 2.5. Observed 90% CL upper limits on the χ-nucleon scattering cross-section as a function of m_χ for the spin-dependent D9 effective operators mediating the interaction of the DM particles with the $q\bar{q}$ initial state. The limits are compared with results from the published ATLAS hadronically decaying W/Z and $j + \chi\chi$ searches, COUPP [40], SIMPLE [41], PICASSO [42], and IceCube [43]. Reproduced with permission from [38]. Copyright 2014 the American Physical Society.

Therefore, a successful particle physics model for DM must comply with all such constraints from the DM observations.

References

[1] Servant G and Tait T M P 2003 Is the lightest Kaluza–Klein particle a viable dark matter candidate? *Nucl. Phys.* **B650** 391–419

[2] Cheng H-C, Feng J L and Matchev K T 2002 Kaluza–Klein dark matter *Phys. Rev. Lett.* **89** 211301

[3] Honorez L L, Nezri E, Oliver J F and Tytgat M H G 2007 The inert doublet model: an archetype for dark matter *J. Cosmol. Astropart. Phys.* **0702** 028

[4] Agrawal P, Dolle E M and Krenke C A 2009 Signals of inert doublet dark matter in neutrino telescopes *Phys. Rev.* **D79** 015015

[5] Nezri E, Tytgat M H G and Vertongen G 2009 e+ and anti-p from inert doublet model dark matter *J. Cosmol. Astropart. Phys.* **0904** 014

[6] Goudelis A, Herrmann B and Stål O 2013 Dark matter in the Inert Doublet Model after the discovery of a Higgs-like boson at the LHC *J. High Energy Phys.* **09** 106

[7] Jungman G, Kamionkowski M and Griest K 1996 Supersymmetric dark matter *Phys. Rep.* **267** 195–373

[8] Bertone G, Hooper D and Silk J 2005 Particle dark matter: evidence, candidates and constraints *Phys. Rep.* **405** 279–390

[9] Bergstrom L 2009 Dark matter candidates *New J. Phys.* **11** 105006

[10] Aprile E 2013 The XENON1T dark matter search experiment *Springer Proc. Phys.* **148** 93–6

[11] Lewin J D and Smith P F 1996 Review of mathematics, numerical factors, and corrections for dark matter experiments based on elastic nuclear recoil *Astropart. Phys.* **6** 87–112

[12] Peter A H G, Gluscevic V, Green A M, Kavanagh B J and Lee S K 2014 WIMP physics with ensembles of direct-detection experiments *Phys. Dark Univ.* **5-6** 45–74

[13] Undagoitia T M and Rauch L 2016 Dark matter direct-detection experiments *J. Phys.* **G43** 013001

[14] Aprile E *et al* 2012 Dark matter results from 225 live days of XENON100 data *Phys. Rev. Lett.* **109** 181301

[15] Akerib D S *et al* 2014 First results from the LUX dark matter experiment at the Sanford Underground Research Facility *Phys. Rev. Lett.* **112** 091303

[16] Bernabei R *et al* 2013 Final model independent result of DAMA/LIBRA-phase1 *Eur. Phys. J.* **C73** 2648

[17] Aalseth C E *et al* 2011 Results from a search for light-mass dark matter with a P-type point contact germanium detector *Phys. Rev. Lett.* **106** 131301

[18] Aalseth C E *et al* 2011 Search for an annual modulation in a P-type point contact germanium dark matter detector *Phys. Rev. Lett.* **107** 141301

[19] Ahmed Z *et al* 2011 Results from a low-energy analysis of the CDMS II germanium data *Phys. Rev. Lett.* **106** 131302

[20] Agnese R *et al* 2013 Silicon detector dark matter results from the final exposure of CDMS II *Phys. Rev. Lett.* **111** 251301

[21] Agnese R *et al* 2014 Search for low-mass weakly interacting massive particles with SuperCDMS *Phys. Rev. Lett.* **112** 241302

[22] Armengaud E *et al* 2012 A search for low-mass WIMPs with EDELWEISS-II heat-and-ionization detectors *Phys. Rev.* **D86** 051701

[23] Bernabei R *et al* 2008 The DAMA/LIBRA apparatus *Nucl. Instrum. Meth.* A **592** 297–315

[24] Aprile E *et al* 2017 First dark matter search results from the XENON1T experiment *Phys. Rev. Lett.* **119** 181301

[25] Akerib D S *et al* 2017 Results from a search for dark matter in the complete LUX exposure *Phys. Rev. Lett.* **118** 021303

[26] Tan A *et al* 2016 Dark matter results from first 98.7 days of data from the PandaX-II experiment *Phys. Rev. Lett.* **117** 121303

[27] Agnese R *et al* 2016 New results from the search for low-mass weakly interacting massive particles with the CDMS low ionization threshold experiment *Phys. Rev. Lett.* **116** 071301

[28] Abe K *et al* 2016 Direct dark matter search by annual modulation in XMASS-I *Phys. Lett.* B **759** 272–76

[29] Adriani O *et al* 2009 An anomalous positron abundance in cosmic rays with energies 1.5-100 GeV *Nature* **458** 607–9

[30] Adriani O *et al* 2009 A new measurement of the antiproton-to-proton flux ratio up to 100 GeV in the cosmic radiation *Phys. Rev. Lett.* **102** 051101

[31] Ackermann M *et al* 2012 Fermi LAT search for dark matter in gamma-ray lines and the inclusive photon spectrum *Phys. Rev.* D **86** 022002

[32] Aguilar M *et al* 2013 First result from the alpha magnetic spectrometer on the International Space Station: precision measurement of the positron fraction in primary cosmic rays of 0.5–350 GeV *Phys. Rev. Lett.* **110** 141102

[33] Abramowski A *et al* 2013 Search for photon line-like signatures from dark matter annihilations with H.E.S.S *Phys. Rev. Lett.* **110** 041301

[34] Bulbul E, Markevitch M, Foster A, Smith R K, Loewenstein M and Randall S W 2014 Detection of an unidentified emission line in the stacked x-ray spectrum of galaxy clusters *Astrophys. J.* **789** 13

[35] Boyarsky A, Ruchayskiy O, Iakubovskyi D and Franse J 2014 Unidentified line in x-ray spectra of the Andromeda Galaxy and Perseus galaxy cluster *Phys. Rev. Lett.* **113** 251301

[36] Urban O, Werner N, Allen S W, Simionescu A, Kaastra J S and Strigari L E 2015 A Suzaku search for dark matter emission lines in the x-ray brightest galaxy clusters *Mon. Not. Roy. Astron. Soc.* **451** 2447–61

[37] Cappelluti N, Bulbul E, Foster A, Natarajan P, Urry M C, Bautz M W, Civano F, Miller E and Smith R K 2018 Searching for the 3.5 keV line in the deep fields with Chandra: the 10 Ms observations *Astrophys. J.* **854** 179

[38] Aad G *et al* 2014 Search for dark matter in events with a Z boson and missing transverse momentum in pp collisions at $\sqrt{s}$=8 TeV with the ATLAS detector *Phys. Rev.* D **90** 012004

[39] Essig R, Mardon J, Papucci M, Volansky T and Zhong Y-M 2013 Constraining light dark matter with low-energy e$^+$ e$^-$ colliders *J. High Energy Phys.* **1311** 167

[40] Behnke E *et al* 2012 First dark matter search results from a 4-kg CF$_3$I bubble chamber operated in a deep underground site *Phys. Rev.* D **86** 052001

[41] Felizardo M *et al* 2012 Final analysis and results of the phase II SIMPLE dark matter search *Phys. Rev. Lett.* **108** 201302

[42] Archambault S *et al* 2012 Constraints on low-mass WIMP interactions on ^{19}F from PICASSO *Phys. Lett.* B **711** 153–61

[43] Aartsen M G *et al* 2013 Search for dark matter annihilations in the Sun with the 79-string IceCube detector *Phys. Rev. Lett.* **110** 131302

[44] Djouadi A, Lebedev O, Mambrini Y and Quevillon J 2012 Implications of LHC searches for Higgs-portal dark matter *Phys. Lett.* B **709** 65–9
[45] Chatrchyan S *et al* 2014 Search for invisible decays of Higgs bosons in the vector boson fusion and associated ZH production modes *Eur. Phys. J.* **C74** 2980
[46] Baltz E A 2004 Dark matter candidates *eConf.* **C040802** L002
[47] Taoso M, Bertone G and Masiero A 2008 Dark matter candidates: a ten-point test *J. Cosmol. Astropart. Phys.* **0803** 022

Chapter 3

WIMP dark matter model: the simplest extension of the Standard Model

As discussed in previous chapters, we need to extend the Standard Model (SM) in order to accommodate a viable dark matter (DM) candidate. One of the simplest ways to accomplish this goal is by extending the particle content of the SM in a minimal way. In this chapter, we will focus on such expansions of the SM.

3.1 Why is WIMP a popular choice?

The most intriguing piece of cosmological evidence in favor of weakly interacting massive particle (WIMP) DM is that thermally produced WIMPs naturally have a relic abundance close to that observed for DM. In the early Universe when the temperature was high enough ($T \gg m_\chi$), the DM particles were in thermal equilibrium with the rest of the cosmic plasma. In order for this particle to remain in thermal equilibrium, it should interact sufficiently with its surroundings. The evolution of the WIMP number density n_χ is given by the Boltzmann equation

$$\frac{dn_\chi}{dt} + 3Hn_\chi = \langle\sigma v\rangle\left(n_{\chi,\,\mathrm{eq}}^2 - n_\chi^2\right), \tag{3.1}$$

where H is the Hubble expansion rate, $\langle\sigma v\rangle$ is the thermally averaged total annihilation cross-section and $n_{\chi,\,\mathrm{eq}}$ is the equilibrium WIMP number density. As the Universe expands and its temperature falls, the number density of WIMPs decreases exponentially. Thus, the annihilation rate reduces and becomes smaller than the Hubble expansion rate. Then the DM species decouples from the cosmic plasma and number density experiences a 'freeze-out'—hence we observe a significant relic abundance of DM today. When freeze-out occurs $\Gamma(T_f) \simeq n(T_f)$ $\langle\sigma v\rangle(T_f) \sim H(T_f)$. Thus, WIMPs freeze out when they are nonrelativistic, at a temperature where their equilibrium number density is Boltzmann suppressed and their velocity is small.

doi:10.1088/978-1-64327-132-3ch3

The WIMP relic abundance is then given by the present-day WIMP density [1],

$$\Omega_\chi h^2 = 1.1 \times 10^9 \frac{x_f}{\sqrt{g^*} \, m_{\text{Pl}} \langle \sigma v \rangle_{\text{ann}}} \text{GeV}^{-1}, \tag{3.2}$$

where $x_f = m_\chi / T_f$, with T_f as the freeze-out temperature. m_{Pl} is Planck mass = 1.22 $\times 10^{19}$ GeV, and, g^* is the effective number of relativistic degrees of freedom (we use, $g^* = 100$ and $x_f = 20$). It can be expressed in a more common form as

$$\Omega_\chi h^2 \approx 0.1 \left(\frac{3 \times 10^{-26} \text{cm}^3 \text{s}^{-1}}{\langle \sigma v \rangle_{\text{ann}}} \right). \tag{3.3}$$

Now, if we calculate the annihilation cross-section of a WIMP, having mass in the weak scale range ($m_\chi \sim M_W$), resulting from a weak interaction, we obtain $\sigma \sim \alpha^2 / m_\chi^2$. Here, α is the fine-structure constant. Thus, we obtain $\sigma \approx 2\text{pb} \left(\frac{100 \text{ GeV}}{m_\chi} \right)^2$. Hence, a weak-scale annihilation cross-section naturally gives a thermally produced WIMP relic abundance that matches the observed DM relic abundance—this striking coincidence is known as the 'WIMP miracle'.

3.2 Singlet scalar dark matter

The scalar singlet extension of the SM [2–7] is the most simplified Higgs-portal model to account for a WIMP candidate. Here, the particle content of the SM is extended by adding a real (gauge) singlet S', which acts as a viable DM candidate. In order to stabilize the DM candidate, we need to impose a $\mathbb{Z}_2$-parity. It interacts only with the SM Higgs boson through the renormalizable interaction term present in the Lagrangian

$$L = L_{\text{SM}} + \frac{1}{2}(')^2 - \frac{1}{2}\mu_{\prime}^{2\prime 2} + L_{\text{int}} - \lambda'^4, \tag{3.4}$$

where $L_{\text{int}} = -\lambda_{\prime}|\Phi|^{2\prime 2}$ denotes the interaction between the SM-Higgs and DM. Since the interaction is mediated only through the Higgs boson, such DM candidates are known as Higgs-portal DM.

After electroweak symmetry breaking (EWSB), the mass of the DM becomes $m_{\text{DM}}^2 = \mu_{\prime}^2 + \frac{1}{2}\lambda_{\prime}v^2$. Once the model framework is set up, we need to analyze and constrain the model parameters from recent evidence of DM. One such major parameter is the coupling between DM and SM-Higgs, i.e. $\lambda_{S'}$, which governs the interaction cross-section and in turn contributes to the relic abundance. $\lambda_{S'}$ is constrained from the invisible decay width of the Higgs boson when $m_{\text{DM}} \lesssim m_h/2$, such that BR $(h \to SS) \lesssim 0.20$. The DM annihilates through SM-Higgs into SM-particles and thus accounts for the correct relic abundance. The relic abundance of DM can be formulated as [1]

$$\Omega_{\text{CDM}} h^2 = 1.1 \times 10^9 \frac{x_f}{\sqrt{g^*} \, m_{\text{Pl}} \langle \sigma v \rangle_{\text{ann}}} \text{GeV}^{-1}, \tag{3.5}$$

where $x_f = m_{DM}/T_D$, with T_D the decoupling temperature, m_{Pl} is Planck mass = 1.22×10^{19} GeV, and $g*$ is the effective number of relativistic degrees of freedom. $\langle \sigma v \rangle_{ann}$ is the thermal averaged value of DM annihilation cross-section times relative velocity. $\langle \sigma v \rangle_{ann}$ can be obtained using the well known formula [8]

$$\langle \sigma v \rangle_{ann} = \frac{1}{m_{DM}^2} \left\{ w(s) - \frac{3}{2}\left(2w(s) - 4m_{DM}^2 w'(s)\right)\frac{1}{x_f} \right\}, \tag{3.6}$$

where prime denotes differentiation with respect to s ($\sqrt{s}$ is the center of mass energy) and evaluated at $s = (2m_{DM})^2$. We have analyzed that such a WIMP candidate cannot produce the required relic-abundance unless a scalar resonance is present, i.e. when, $m_{S'} \simeq m_h/2$ GeV, when the annihilation cross-section is enhanced. The coupling and scattering cross-section can also be constrained from direct detection experiments. Thus, one can scan the parameter space to find out the allowed region.

3.3 Singlet fermionic dark matter

The singlet fermionic dark matter (SFDM) model is a renormalizable extension of the SM with a hidden sector containing a scalar singlet Φ_s and a singlet Dirac fermion ψ [9–14]. Here, the SFDM ψ interacts with the SM sector via the singlet Φ_s, which mixes with the SM-Higgs doublet Φ. Therefore, this is also an example of a Higgs-portal model. The Lagrangian of the SFDM model is given as

$$\mathcal{L} = \mathcal{L}_{SM} + \mathcal{L}_{hid} + \mathcal{L}_{int}, \tag{3.7}$$

where

$$\mathcal{L}_{hid} = \mathcal{L}_{\Phi_s} + \bar{\psi}(i\partial_\mu \gamma^\mu - m_\psi)\psi - \lambda_{\psi S}\,\bar{\psi}\psi\,\Phi_s, \tag{3.8}$$

$$\mathcal{L}_{int} = \frac{\lambda_1'}{2}\Phi^\dagger \Phi \Phi_s + \frac{\lambda_2'}{2}\Phi^\dagger \Phi \Phi_s^2 \tag{3.9}$$

and

$$\mathcal{L}_{\Phi_s} = \frac{1}{2}(\partial \Phi_s)^2 - \frac{m_{\Phi_s}^2}{2}\Phi_s^2 - \frac{\lambda'}{3}\Phi_s^3 - \frac{\lambda''}{4}\Phi_s^4. \tag{3.10}$$

After EWSB, the singlet field Φ_s can be written as $\Phi_s = x + s$, where x is the vacuum expectation value (VEV) of Φ_s. The two scalar eigenstates are denoted as

$$h_2 = \sin\theta\, s + \cos\theta\, \phi, \tag{3.11}$$

$$h_1 = \sin\theta\, \phi - \cos\theta\, s, \tag{3.12}$$

where h_2 is identified as the SM-Higgs boson and $\cos\theta(\sin\theta)$ is the scalar-mixing. Now, the mass of the DM is given by $m_{DM} = m_\psi + \lambda_{\psi S}\, x$, with m_ψ as a free parameter. The DM interaction strength depends on the parameter $\lambda_{DM} = \lambda_{\psi S}$. Here, the scalar mixing angle and DM-coupling are subject to various constraints, such as the Large Hadron Collider (LHC) bound on the SM-Higgs boson, the relic abundance of DM and the upper bound on the DM-nucleon scattering cross-section.

3.3.1 Constraints from LHC

Observation of the SM-like Higgs boson at LHC by the CMS [15] and ATLAS [16] collaborations will constrain this mixing angle severely. The signal strength or reduction factor of a particular channel can be defined as

$$r_i^{xx} = \frac{\sigma_{H_i}}{\sigma_{H_i}^{SM}} \cdot \frac{BR_{H_i \to xx}}{BR_{H_i \to xx}^{SM}}, \quad (i = 1, 2). \tag{3.13}$$

where, σ_{H_i} and $BR_{H_i \to xx}$ are the production cross-section of H_i, and the branching ratio of $H_i \to xx$, respectively. Similarly, $\sigma_{H_i}^{SM}$ and $BR_{H_i \to xx}^{SM}$ are the corresponding quantities of the SM-Higgs. Using equation (3.13) one obtains

$$
\begin{aligned}
r_2 &= \cos^4 \alpha \frac{\Gamma_{H_2}^{SM}}{\cos^2 \alpha \; \Gamma_{H_2}^{SM} + \sin^2 \alpha \; \Gamma_{H_2}^{Hid} + \Gamma_{H_2 \to H_1 H_1}} \\
r_1 &= \sin^4 \alpha \frac{\Gamma_{H_1}^{SM}}{\sin^2 \alpha \; \Gamma_{H_1}^{SM} + \cos^2 \alpha \; \Gamma_{H_1}^{Hid}}
\end{aligned}
\tag{3.14}
$$

where $\Gamma_{H_i}^{SM}$ denotes the total decay width of the SM-Higgs boson and $\Gamma_{H_i}^{Hid}$ is the invisible decay width ($H_i \to 2$ DM). The invisible decay width of the SM Higgs reads as

$$\Gamma_{H_2}^{Hid} \equiv \Gamma_{inv} = \frac{m_{H_2} \lambda_{DM}^2}{16\pi} \sin^2 \alpha \left(1 - 4\frac{m_{DM}^2}{m_{H_2}^2} \right)^{\frac{3}{2}}. \tag{3.15}$$

For $m_{DM} < m_{H_2}/2$, we can constrain the DM coupling λ_{DM} from the invisible decay width of SM-Higgs boson. Figure 3.1 shows the allowed range of λ_{DM} with mass of DM for different invisible branching ratios of the SM-Higgs boson, assuming the width of the Higgs to SM fermions as 4.21 MeV. As an example, if we consider $m_{DM} \sim 30$ GeV, if $BR_{inv} \geqslant 20\%$ (35%) then DM-coupling, λ_{DM}, should be less than 0.06 (0.075). Again, the signal strength (as defined in equations (3.13)–(3.14)) depends on the scalar mixing angle. Constraining r_2 to be $\leqslant 0.9$ (or 0.8), we obtain the allowed range of scalar mixing $\cos \alpha$ as a function of m_{DM} for a particular value of DM-coupling.

3.3.2 Constraints from relic density and direct detection

We obtain the relic abundance (using equation (3.5)) of the DM in agreement with the WMAP-9 year result [17] and PLANCK [18], only near resonance where $m_{DM} = m_{h_1}/2$ GeV. Dominant contribution to relic density comes from final-state $b\bar{b}$ annihilation with cross-section $\langle \sigma v \rangle \simeq 1.7 \times 10^{-26} \text{cm}^3 \text{s}^{-1}$. We observe that as we

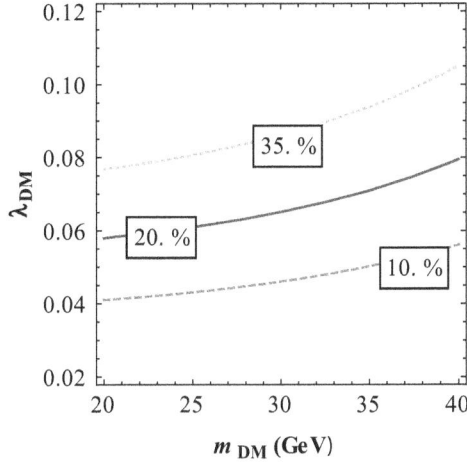

Figure 3.1. Contours of invisible branching ratio for SFDM model (10%, 20% and 35%) in the plane of $[m_{DM}, \lambda_{DM}]$ with $\cos \alpha = 0.95$.

decrease λ_{DM}, the annihilation cross-section is also decreased. But, if we approach very near the resonance region, i.e. $m_{H_1} - 2m_{DM} \sim \mathcal{O}(10^{-4})$, the annihilation cross-section can be enhanced significantly, which counterbalances the previous effect. However, if we are slightly away from resonance we need to have $\lambda_{DM} \sim 10^{-2}$ to obain the correct relic.

The scattering cross-section (spin-independent) for the DM off a proton or neutron is

$$\sigma_{p,n}^{SI} = \frac{4m_r^2}{\pi} f_{p,n}^2, \tag{3.16}$$

where m_r is the reduced mass, defined as $1/m_r = 1/m_{DM} + 1/m_{p,n}$ and $f_{p,n}$ is the hadronic matrix element, given by

$$f_{p,n} = \sum_{q=u,d,s} f_{Tq}^{(p,n)} a_q \frac{m_{p,n}}{m_q} + \frac{2}{27} f_{TG}^{(p,n)} \sum_{q=c,b,t} a_q \frac{m_{p,n}}{m_q}.$$

The f-values are given in [19]. Here, a_q is the effective coupling constant between the DM and the quark. An approximate form of a_q/m_q can be recast as:

$$\frac{a_q}{m_q} = \frac{\lambda_{DM}}{v\sqrt{2}} \left[\frac{1}{m_h^2} - \frac{1}{m_H^2} \right] \sin \alpha \cos \alpha. \tag{3.17}$$

In order to be consistent with the latest exclusion limit on σ_p^{SI} as specified by LUX [20] and XENON 100 [21, 22], we require $\sigma_p^{SI} \lesssim 10^{-45} \text{cm}^2$. In figure 3.2, we show the contour of $\sigma_p^{SI} = 10^{-45} \text{cm}^2$ (red solid line). It indicates that λ_{DM} should be small enough (in the range of $\sim 10^{-4}$–10^{-5}) to satisfy the required value of σ_p^{SI}. As argued before, the very near resonance region, for $\lambda_{DM} \sim 10^{-4}$, also gives correct relic

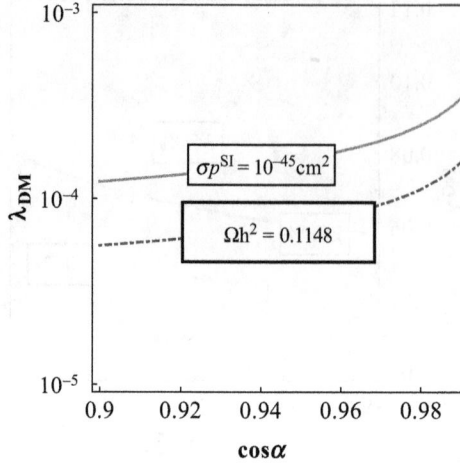

Figure 3.2. Contours of relic abundance (blue dot-dashed line) consistent with WMAP9 [17] and spin-independent scattering cross-sections for the SFDM model, $\sigma_p^{\text{SI}} = 10^{-45}$ cm^2 (red solid line), in the plane of $[\lambda_{\text{DM}}, \cos \alpha]$ for $m_{\text{DM}} \sim 31$ GeV.

density. The contour of relic abundance is shown in figure 3.2 by the blue dot-dashed line.

3.4 Inert doublet Higgs model

The inert doublet model (IDM) is one of the simplest possible extensions of the SM, with two Higgs doublets (for a literature review see [23–26]), $\Phi_{\text{S,D}}$. Φ_{S} is completely analogous to the SM-Higgs doublet and Φ_{D} is the inert doublet, i.e. it does not interact with SM fermions. The renormalizable $SU(2) \times U(1)$ invariant Higgs potential is given by

$$V = \mu_1^2 A + \mu_2^2 B + \lambda_1 A^2 + \lambda_2 B^2 + \lambda_3 AB + \lambda_4 C^\dagger C + \frac{\lambda_5}{2}(C^2 + C^{\dagger 2}), \qquad (3.18)$$

where A, B, C are given by

$$A = \Phi_{\text{S}}^\dagger \Phi_{\text{S}}, \ B = \Phi_{\text{D}}^\dagger \Phi_{\text{D}}, \ C = \Phi_{\text{S}}^\dagger \Phi_{\text{D}} \quad . \qquad (3.19)$$

After spontaneous symmetry breaking

$$\langle \Phi_{\text{S}} \rangle = \frac{1}{\sqrt{2}} \begin{pmatrix} 0 \\ v \end{pmatrix}, \quad \langle \Phi_{\text{D}} \rangle = \begin{pmatrix} 0 \\ 0 \end{pmatrix} \quad , \qquad (3.20)$$

where $\langle \Phi_{\text{D}} \rangle$ is forced by the discrete symmetry, and $v = 246$ GeV. Expanding the fields around vacua,

$$\Phi_{\text{S}} = \begin{pmatrix} G^+ \\ (v + h + iG^0)/\sqrt{2} \end{pmatrix}, \quad \Phi_{\text{D}} = \begin{pmatrix} D^+ \\ (H + iA)/\sqrt{2} \end{pmatrix}, \qquad (3.21)$$

where G^0 and $G^\pm$ are the neutral and charged Goldstone bosons, and h is the SM-like Higgs boson. The fields in the second doublet are:

- scalar H and pseudoscalar A, both neutral $\rightarrow$ dark sector,
- charged scalar $D^\pm$ $\rightarrow$ inert sector.

The masses of the physical states are

$$m_h^2 = 2\lambda_1 v^2$$
$$m_D^2 = \mu_2^2 + \frac{\lambda_3}{2} v^2$$
$$m_H^2 = m_D^2 + \left(\frac{\lambda_4 + \lambda_5}{2}\right) v^2 = \mu_2^2 + \frac{\lambda_{345}}{2} v^2 \qquad (3.22)$$
$$m_A^2 = m_D^2 + \left(\frac{\lambda_4 - \lambda_5}{2}\right) v^2$$

with $\lambda_{345} = \lambda_3 + \lambda_4 + \lambda_5$.

3.4.1 Dark matter in IDM

There are two possible DM candidates—either the scalar H or the pseudoscalar A, of Φ_D. The discrete symmetry prevents the direct coupling of Φ_D to fermions, and hence guarantees the stability of the lightest inert particle. Let us choose, $m_H^2 < m_A^2$. Hence, we consider H to be the DM candidate. The parameter λ_{345} is related to triple and quartic couplings between the SM-like Higgs h and the DM candidate H. λ_2 gives the quartic DM self-couplings. Relevant LHC data constrain the parameters of IDM as follows:

- Higgs mass $M_h = 125.09 \pm 0.24$ GeV.
- Higgs total decay width $\Gamma_{\text{tot}} = (4.2 - 5.5) \times 4.5$ MeV.
- Higgs total signal strength $\mathcal{R} = 1.09 \pm 0.11$.
- Higgs decay into $\gamma\gamma$ signal strength $\mathcal{R}_{\gamma\gamma} = 1.16^{+0.20}_{-0.18}$.
- Invisible decay branching ratio $\text{BR}(h \rightarrow \text{inv}) < 0.23 - 0.36$.

The Higgs boson of the IDM, apart from the SM decay channels, has additional channels, such as deay into a pair of dark particles, AA, HH or $H^\pm H^\mp$. The decay width for the process $h \rightarrow HH$ is

$$\Gamma(h \rightarrow HH) = \frac{\lambda_{345}^2 v^2}{32\pi M_h} \sqrt{1 - \frac{4M_H^2}{M_h^2}}. \qquad (3.23)$$

The parameter space of this IDM can be restricted taking into account all such constraints.

The relic abundance can be calculated using the micrOMEGAs software [27, 28]. A case study (as analyzed in the literature [26]) of a random scan of parameters with relic abundance as a constraint is shown in figure 3.3. The figure shows the plot of relic abundance due to DM candidate H versus the mass of DM. It shows there are

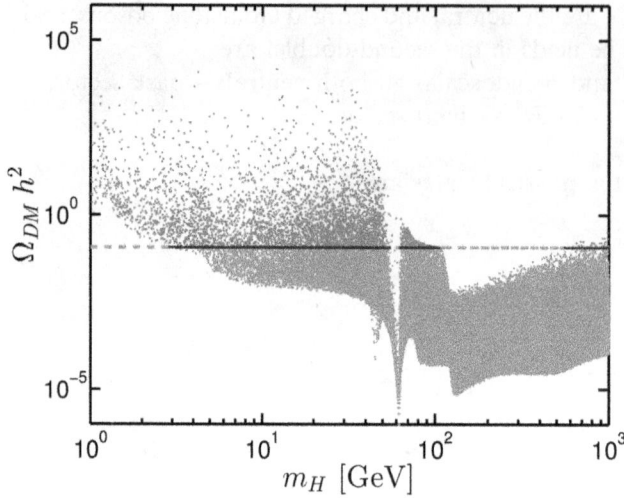

Figure 3.3. Plot of relic abundance versus the mass of DM consistent with $\Omega_{\rm DM}h^2 = 0.1181 \mp 0.0012$. Reproduced with permission from [26]. Copyright 2016 Hindawi.

three distinct mass regions (low, moderate and high) separated by two mass-gaps, where the observed relic, i.e. $\Omega_{\rm DM}h^2 = 0.1181 \pm 0.0012$ is satisfied. The first mass gap appears around $m_H \approx m_h/2$, i.e. the resonance region. The second one appears near $m_H \approx m_{W,\,Z}$, thereby the annihilation channels into WW or ZZ final states open up. For the high mass region the annihilation process $HH \rightarrow hh$ plays the dominant role in obtaining the relic abundance.

3.5 Two-component dark matter model

The two-component DM model [29–35] is a widely studied extension of the SM, where two DM candidates contribute to make up the observed relic abundance. There are several variants of this model proposed in the literature, such as both candidates being scalars, or one being scalar and one fermionic. There is also interaction between the two DM candidates, of which one may be the decaying DM. Here we discuss a simple scalar–fermion two-component model.

3.5.1 The model

In this case, let us consider an extension of the SM by two gauge-singlet scalars (ϕ_1 and ϕ_2) and one fermion. The fermion (ψ) and one scalar (ϕ_1) are assumed to be odd under a Z_2 symmetry to ensure the stability of the lightest odd particle. The other scalar (ϕ_2) is even under Z_2. Thus, the model contains two DM particles.

The mass and interaction terms involving the DM fermion (ψ) are given by

$$\mathcal{L} = -\frac{1}{2}(M_\psi \bar{\psi}\psi + g_1\phi_2\bar{\psi}\psi + g_2\phi_2\bar{\psi}\gamma_5\psi), \tag{3.24}$$

where g_1, g_2 are, respectively, the scalar and pseudoscalar couplings of ψ. There are no interaction terms involving both ψ and ϕ_1, since it is not invariant under the gauge and the Z_2 symmetry.

The scalar potential of this model can be written as

$$V(\phi_1, \phi_2, h) = V_h + V_{\phi_1} + V_{\phi_2} + V_{\text{int}}, \tag{3.25}$$

where ϕ_1 is the scalar DM particle and h is the SM-Higgs doublet. h (the SM-Higgs doublet) and ϕ_2 mix with each other giving rise to two scalar mass eigenstates, H_1 and H_2, defined as

$$H_1 = h \cos \alpha + \phi_2 \sin \alpha, \quad H_2 = \phi_2 \cos \alpha - h \sin \alpha, \tag{3.26}$$

where α is the mixing angle. We assume this mixing angle to be small so that we can identify H_1 with the SM-like Higgs observed at the LHC [15, 16] with a mass of about 125 GeV. The other scalar, H_2, can be considered to be heavier than H_1.

3.5.2 Relic density

Since the model contains two DM particles, ψ and ϕ_1, we need to simultaneously follow their abundances in the early Universe,

$$\Omega_{\text{DM}}h^2 = \Omega_\psi h^2 + \Omega_{\phi_1} h^2. \tag{3.27}$$

The Boltzmann equations [36] have to be solved for each of the species. Since, both DM candidates interact through the Higgs-portal, now there arises the possibility of processes such as $\psi\psi \to \phi_1\phi_1$ and vice versa. In this process, one DM candidate can be converted into another—this would affect the relic abundance of each of the species. As we have analyzed the nature of Higgs portal DM, it has been observed that the correct relic abundance is obtained at the resonance point where $m_{\text{DM}} \simeq m_{H_{1,2}}/2$. Here, one can study the detailed parameter space of such models in order to obtain correct relic abundance.

References

[1] Kolb E W and Turner M S 1990 The early universe *Front. Phys.* **69** 1–547

[2] McDonald J 1994 Gauge singlet scalars as cold dark matter *Phys. Rev.* D **50** 3637–49

[3] Burgess C P, Pospelov M and ter Veldhuis T 2001 The minimal model of nonbaryonic dark matter: a singlet scalar *Nucl. Phys.* B **619** 709–28

[4] Davoudiasl H, Kitano R, Li T and Murayama H 2005 The new minimal Standard Model *Phys. Lett.* B **609** 117–23

[5] Bandyopadhyay A, Chakraborty S, Ghosal A and Majumdar D 2010 Constraining scalar singlet dark matter with CDMS, XENON and DAMA and prediction for direct detection rates *J. High Energy Phys.* **1011** 065

[6] Guo W-L and Wu Y-L 2010 The real singlet scalar dark matter model *J. High Energy Phys.* **1010** 083

[7] Cline J M, Kainulainen K, Scott P and Weniger C 2013 Update on scalar singlet dark matter *Phys. Rev.* D **88** 055025 Cline J M, Kainulainen K, Scott P and Weniger C 2015 (Erratum *Phys. Rev.* D **92** 039906)

[8] Srednicki M, Watkins R and Olive K A 1988 Calculations of relic densities in the early Universe *Nucl. Phys.* B **310** 693

[9] Kim Y G, Lee K Y and Shin S 2008 Singlet fermionic dark matter *J. High Energy Phys.* **0805** 100

[10] Kim Y G and Shin S 2009 Singlet fermionic dark matter explains DAMA signal *J. High Energy Phys.* **0905** 036

[11] Baek S, Ko P and Park W-I 2012 Search for the Higgs portal to a singlet fermionic dark matter at the LHC *J. High Energy Phys.* **1202** 047

[12] Baek S, Ko P, Park W-I and Senaha E 2012 Vacuum structure and stability of a singlet fermion dark matter model with a singlet scalar messenger *J. High Energy Phys.* **1211** 116

[13] Esch S, Klasen M and Yaguna C E 2013 Detection prospects of singlet fermionic dark matter *Phys. Rev.* D **88** 075017

[14] Ettefaghi M M and Moazzemi R 2013 Annihilation of singlet fermionic dark matter into two photons *J. Cosmol. Astropart. Phys.* **1302** 048

[15] Chatrchyan S *et al* 2012 Observation of a new boson at a mass of 125 GeV with the CMS experiment at the LHC *Phys. Lett.* B **716** 30–61

[16] Aad G *et al* 2012 Observation of a new particle in the search for the Standard Model Higgs boson with the ATLAS detector at the LHC *Phys. Lett.* B **716** 1–29

[17] Hinshaw G *et al* 2013 Nine-year Wilkinson microwave anisotropy probe (WMAP) observations: cosmological parameter results *Astrophys. J. Suppl.* **208** 19

[18] Ade P A R *et al* 2013 Planck 2013 results. XVI. Cosmological parameters *Astron. Astrophys.* **571** A16

[19] Ellis J R, Ferstl A and Olive K A 2000 Re-evaluation of the elastic scattering of supersymmetric dark matter *Phys. Lett.* B **481** 304–14

[20] Akerib D S *et al* 2014 First results from the LUX dark matter experiment at the Sanford Underground Research Facility *Phys. Rev. Lett.* **112** 091303

[21] Baudis L 2012 Results from the XENON100 Dark Matter Search Experiment, arXiv:1203.1589

[22] Aprile E *et al* 2012 Dark matter results from 225 live days of XENON100 data *Phys. Rev. Lett.* **109** 181301

[23] Honorez L L and Yaguna C E 2010 The inert doublet model of dark matter revisited *J. High Energy Phys.* **09** 046

[24] Goudelis A, Herrmann B and Stål O 2013 Dark matter in the inert doublet model after the discovery of a Higgs-like boson at the LHC *J. High Energy Phys.* **09** 106

[25] Arhrib A, Sming Tsai Y-L, Yuan Q and Yuan T-C 2014 An updated analysis of inert Higgs doublet model in light of the recent results from LUX, PLANCK, AMS-02 and LHC *J. Cosmol. Astropart. Phys.* **1406** 030

[26] Díaz M A, Koch B and Urrutia-Quiroga S 2016 Constraints to dark matter from inert Higgs doublet model *Adv. High Energy Phys.* **2016** 8278375

[27] Belanger G, Boudjema F, Pukhov A and Semenov A 2014 micrOMEGAs 3: a program for calculating dark matter observables *Comput. Phys. Commun.* **185** 960–85

[28] Belanger G, Boudjema F, Pukhov A and Semenov A 2010 micrOMEGAs: a tool for dark matter studies *Nuovo Cim.* **C033N2** 111–6

[29] Fairbairn M and Zupan J 2009 Dark matter with a late decaying dark partner *J. Cosmol. Astropart. Phys.* **0907** 001

[30] Biswas A, Majumdar D, Sil A and Bhattacharjee P 2013 Two component dark matter: a possible explanation of 130 GeV γ-ray line from the Galactic Centre *J. Cosmol. Astropart. Phys.* **1312** 049

[31] Bian L, Ding R and Zhu B 2014 Two component Higgs-portal dark matter *Phys. Lett.* B **728** 105–13

[32] Bhattacharya S, Drozd A, Grzadkowski B and Wudka J 2013 Two-component dark matter *J. High Energy Phys.* **10** 158

[33] Biswas A 2016 Explaining low energy γ-ray excess from the Galactic Centre using a two component dark matter model *J. Phys.* G **43** 055201

[34] Biswas A, Majumdar D and Roy P 2015 Nonthermal two component dark matter model for Fermi-LAT-ray excess and 3.55 keV x-ray line *J. High Energy Phys.* **04** 065

[35] Esch S, Klasen M and Yaguna C E 2014 A minimal model for two-component dark matter *J. High Energy Phys.* **09** 108

[36] Belanger G, Kannike K, Pukhov A and Raidal M 2012 Impact of semi-annihilations on dark matter phenomenology—an example of Z_N symmetric scalar dark matter *J. Cosmol. Astropart. Phys.* **1204** 010

Chapter 4

Generation of neutrino mass and dark matter: a unified scenario

The most unambiguous evidence for studying new physics is from the experimental observation of dark matter (DM) and neutrino oscillation. These are the two fundamental issues which cannot be addressed in the realm of the Standard Model (SM). An interesting question is whether these two hints of physics beyond the Standard Model (BSM) could be related in some way. There have already been a lot of studies in this direction. In this context, we will focus on a few such models.

4.1 The see-saw mechanism and dark matter

The SM predicts that neutrinos are massless. The model does not predict that neutrinos have a non-zero mass and mix among the neutrinos in different families. However, neutrino oscillation experiments suggest that neutrinos of a particular flavor can transmute into another flavor. This phenomenon of neutrinos 'oscillating' from one flavor to another is only possible if different neutrino types have different masses. Therefore, neutrino oscillation experiments strongly indicate that neutrinos have mass. This provides a smoking gun signature of new physics BSM.

An attractive mechanism for explaining small neutrino masses is the so-called see-saw mechanism. In the well known canonical see-saw mechanism [1, 2], three heavy gauge-singlet Majorana neutrinos N_i ($i = 1, 2, 3$) are added to the SM of elementary particles, so that

$$\mathcal{M}_\nu^{(e, \mu, \tau)} = -\mathcal{M}_D \mathcal{M}_N^{-1} \mathcal{M}_D^T, \tag{4.1}$$

where $\mathcal{M}_D$ is the 3×3 Dirac mass matrix linking the observed neutrinos ν_α ($\alpha = e, \mu, \tau$) to N_i, and $\mathcal{M}_N$ is the Majorana mass matrix of N_i, which can be much larger than the electroweak (EW) scale. Thus, the neutrino mass turns out to

doi:10.1088/978-1-64327-132-3ch4

be $\mathcal{M}_\nu = \frac{\mathcal{M}_D^2}{\mathcal{M}_N}$. In such cases, the right-handed (RH) neutrino can be a viable DM candidate if it is stabilized by an additional Z_2-symmetry.

4.2 Gauged $B - L$ extension of the Standard Model

The minimal gauge extension of the SM with $U(1)_{B - L}$, and a discrete symmetry (Z_2-parity) has been studied by several authors [2–7] in the context of DM and neutrino mass generation. Along with the SM particles, this model contains an SM singlet S with $B - L$ charge $+2$ and three RH neutrinos N_R^i ($i = 1, 2, 3$) with $B - L$ charge -1. As this $U(1)_{B - L}$ symmetry is gauged, an extra gauge boson Z' is associated as a signature of the extended symmetry. Once the $B - L$ symmetry is broken spontaneously through the vacuum expectation value (VEV) of S, this Z' becomes massive. Here, we also impose a Z_2 discrete symmetry. We assign Z_2 charge $+1$ (or even) for all the particles except N_R^3 [3, 8]. This ensures the stability of N_R^3 which qualified as a viable DM candidate. The assignment of $B - L$ charge in this model eliminates the triangular $B - L$ gauge anomalies and ensures the gauge invariance of the theory. The particle content of this model is shown in table 4.1.

The scalar Lagrangian of this model can be written as

$$\mathcal{L}_s = (D^\mu \Phi)^\dagger D_\mu \Phi + (D^\mu S)^\dagger D_\mu S - V(\Phi, S), \tag{4.2}$$

where the potential term is

$$V(\Phi, S) = m^2 \Phi^\dagger \Phi + \mu^2 |S|^2 + \lambda_1 (\Phi^\dagger \Phi)^2 + \lambda_2 |S|^4 + \lambda_3 \Phi^\dagger \Phi |S|^2, \tag{4.3}$$

with Φ and S as the Higgs doublet and singlet fields, respectively. After spontaneous symmetry breaking (SSB) the two scalar fields take the VEV v and v_{B-L} respectively. Using the minimization conditions, we have the following scalar mass matrix:

$$\mathcal{M} = \begin{pmatrix} \lambda_1 v^2 & \dfrac{\lambda_3 v_{B-L} v}{2} \\ \dfrac{\lambda_3 v_{B-L} v}{2} & \lambda_2 v_{B-L}^2 \end{pmatrix} = \begin{pmatrix} \mathcal{M}_{11} & \mathcal{M}_{12} \\ \mathcal{M}_{21} & \mathcal{M}_{22} \end{pmatrix}. \tag{4.4}$$

The expressions for the scalar mass eigenvalues ($m_H > m_h$) are:

Table 4.1. Particle content of minimal $U(1)_{B - L}$ model.

Particle	Q	u_R	d_R	L	e_R	Φ	S	$N_R^{1,2}$	N_R^3
$SU(2)_L$	2	1	1	2	1	2	1	1	1
$U(1)_Y$	1/6	2/3	−1/3	−1	−1	1	0	0	0
$U(1)_{B - L}$	1/3	1/3	1/3	−1	−1	0	2	−1	−1
Z_2	+	+	+	+	+	+	+	+	−

$$m_{H,h}^2 = \frac{1}{2}\left[\mathcal{M}_{11} + \mathcal{M}_{22} \pm \sqrt{(\mathcal{M}_{11} - \mathcal{M}_{22})^2 + 4\mathcal{M}_{12}^2}\right]. \tag{4.5}$$

The mass eigenstates are linear combinations of ϕ and ϕ', and written as

$$\begin{pmatrix} h \\ H \end{pmatrix} = \begin{pmatrix} \cos\alpha & -\sin\alpha \\ \sin\alpha & \cos\alpha \end{pmatrix}\begin{pmatrix} \phi \\ \phi' \end{pmatrix}, \tag{4.6}$$

where h is the SM-like Higgs boson. The scalar mixing angle α can be expressed as

$$\tan(2\alpha) = \frac{2\mathcal{M}_{12}}{\mathcal{M}_{11} - \mathcal{M}_{22}} = \frac{\lambda_3 v_{B-L} v}{\lambda_1 v^2 - \lambda_2 v_{B-L}^2}. \tag{4.7}$$

Now we can calculate the quartic coupling constants by using equations (4.5), (4.6) and (4.7):

$$\begin{aligned}
\lambda_1 &= \frac{m_H^2}{4v^2}(1 - \cos 2\alpha) + \frac{m_h^2}{4v^2}(1 + \cos 2\alpha), \\
\lambda_2 &= \frac{m_h^2}{4v_{B-L}^2}(1 - \cos 2\alpha) + \frac{m_H^2}{4v_{B-L}^2}(1 + \cos 2\alpha), \\
\lambda_3 &= \sin 2\alpha\left(\frac{m_H^2 - m_h^2}{2\,vv_{B-L}}\right).
\end{aligned} \tag{4.8}$$

In the presence of an extra $U(1)_{B-L}$ gauge theory the SM gauge kinetic term is modified by

$$\mathcal{L}_{B-L}^{K.E} = -\frac{1}{4}F'^{\mu\nu}F'_{\mu\nu}, \tag{4.9}$$

where

$$F'_{\mu\nu} = \partial_\mu B'_\nu - \partial_\nu B'_\mu. \tag{4.10}$$

Here, we consider only the 'pure' $B - L$ model, that is defined by the condition $\tilde{g} = 0$ at the EW scale. This implies zero mixing at tree level between the Z' and Z bosons.

4.2.1 Neutrino mass generation

The relevant Yukawa coupling to generate neutrino masses via a type-I see-saw mechanism is given by

$$\mathcal{L}_{\text{int}} = \sum_{\beta=1}^{3}\sum_{j=1}^{2} y_\beta^j \overline{l}_\beta \tilde{\Phi} N_j - \sum_{i=1}^{3}\frac{y_{n_i}}{2}\overline{N_R^i}SN_R^i, \tag{4.11}$$

where $\tilde{\Phi} = -i\tau_2\Phi^*$.

The neutrino mass can be generated in this model via the type-I see-saw mechanism, where the mass matrices for light and heavy neutrino are given as

$$m_{\nu_L} \simeq m_D^T m_M^{-1} m_D, \tag{4.12}$$

$$m_{\nu_H} \simeq m_M, \tag{4.13}$$

where $m_D = (y_\beta^j/\sqrt{2})v$, $(j = 1, 2)$ and $m_{M_i} = -(y_{n_i}/\sqrt{2})v_{B-L}$, $(i = 1, 2, 3)$.

Because of Z_2-parity, N_R^3 has no Yukawa coupling with the left-handed lepton doublet, therefore the lightest neutrino remains massless. The masses of N_R^1 and N_R^2 are considered to be heavier than that of N_R^3.

4.2.2 Calculation of relic abundance

The relic abundance of DM can be formulated as [9]

$$\Omega_{CDM} h^2 = 1.1 \times 10^9 \frac{x_f}{\sqrt{g^*} m_{Pl} \langle \sigma v \rangle_{ann}} \text{GeV}^{-1}, \tag{4.14}$$

where $x_f = m_{N_R^3}/T_D$, with T_D the decoupling temperature. m_{Pl} is Planck mass $= 1.22 \times 10^{19}$ GeV, and g^* is the effective number of relativistic degrees of freedom (we use, $g^* = 100$ and $x_f = (1/20)$). $\langle \sigma v \rangle_{ann}$ is the thermal averaged value of DM annihilation cross-section times relative velocity. DM interacts with the SM particles via the Z'-boson and h, H. However, as the Z'-boson is heavy ($m_{Z'} \geqslant 2.33$ TeV [10]), the annihilation of DM into the SM particles takes place via h and H only. Thus, effectively we obtain a Higgs-portal DM model.

$\langle \sigma v \rangle_{ann}$ can be obtained using the well known formula [11]

$$\langle \sigma v \rangle_{ann} = \frac{1}{m_{N_R^3}^2} \left\{ w(s) - \frac{3}{2}\left(2w(s) - 4m_{N_R^3}^2 w'(s)\right)\frac{1}{x_f} \right\}\Bigg|_{s=\left(2m_{N_R^3}\right)^2}, \tag{4.15}$$

where prime denotes differentiation with respect to s ($\sqrt{s}$ is the center of mass energy). Here, the function $w(s)$ (detailed calculation in appendix A.1) depends on the amplitude of different annihilation processes,

$$N_R^3 N_R^3 \longrightarrow b\bar{b}, \; \tau^+\tau^-, \; W^+W^-, \; ZZ, \; hh. \tag{4.16}$$

In this model, the RH neutrino N_R^3 turns out to be a viable DM candidate as an artifact of the Z_2 charge assignment. We choose a specific set of benchmark values

Table 4.2. Choice of parameters.

m_h	Γ_h	v_{B-L}	g_{B-L}
125 GeV	4.7×10^{-3} GeV	7 TeV	0.1

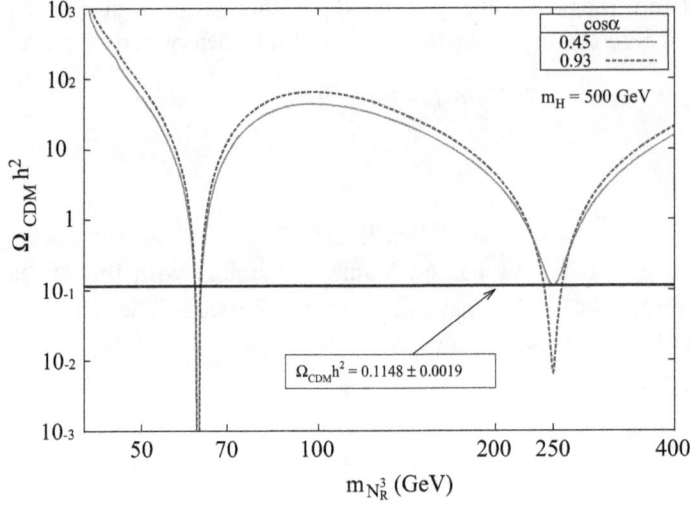

Figure 4.1. Plot of relic abundance as a function of DM mass for $m_H = 500$ GeV with specific choices of scalar mixing angle $\cos \alpha = 0.935$ (blue dashed line), 0.45 (red solid line). The straight line shows the WMAP-9 value, $\Omega_{CDM}h^2 = 0.1148 \pm 0.0019$.

(mass, m_h, and decay width, Γ_h, of the SM-like Higgs boson, VEV of singlet scalar S and $U(1)_{B-L}$ gauge coupling) for our calculation, shown in table 4.2, based on present experimental constraints [10]. However, the mass of the heavy scalar and the scalar mixing angle are not fixed.

In figure 4.1 the relic density is plotted against DM mass for two specific choices (to be explained later in this section) of scalar mixing angles $\cos \alpha = 0.935, 0.45$ with $m_H = 500$ GeV. The straight line shows the latest nine-year WMAP data, i.e. $\Omega_{CDM}h^2 = 0.1148 \pm 0.0019$ [12] (whereas the latest PLANCK result is $\Omega_{CDM}h^2 = 0.1199 \pm 0.0027$ at 68% CL [13]). The resultant relic abundance is found to be consistent with the reported value of the WMAP-9 and PLANCK experiments only near resonance when $m_{N_R^3} \sim (1/2)m_{h,\,H}$[1]. The reason for the over abundance of DM except at resonance can be understood in the following way: the annihilation cross section of DM, being proportional to $y_{n_3}^2$ (where, $y_{n_3} = (\sqrt{2}\,m_{N_R^3})/v_{B-L}$), is heavily suppressed due to the large value of v_{B-L}.

In future, PLANCK data could further restrict the choice of parameter space. The total annihilation cross-section is enhanced due to scalar resonance, otherwise it will be suppressed due to heavy Z'.

4.2.3 Spin-independent scattering cross-section

The effective Lagrangian describing the elastic scattering of the DM off a nucleon is given by

---

[1] In principle, Z' resonance can also provide the correct relic abundance, but in that case the DM mass will be $\mathcal{O}$ (TeV) (i.e. $m_{N_R^3} \sim (1/2)\,m_{Z'}$), if we consider the current experimental bound on Z' mass [10].

$$L_{\text{eff}} = f_p \bar{N}_R^3 N_R^3 \bar{p} p + f_n \bar{N}_R^3 N_R^3 \bar{n} n, \tag{4.17}$$

where $f_{p,\,n}$ is the hadronic matrix element, given by

$$f_{p,\,n} = \sum_{q=u,d,s} f_{Tq}^{(p,\,n)} a_q \frac{m_{p,\,n}}{m_q} + \frac{2}{27} f_{TG}^{(p,\,n)} \sum_{q=c,b,t} a_q \frac{m_{p,\,n}}{m_q}. \tag{4.18}$$

The f-values are given as in [14]

$$f_{Tu}^{(p)} = 0.020 \pm 0.004, \quad f_{Td}^{(p)} = 0.026 \pm 0.005, \quad f_{Ts}^{(p)} = 0.118 \pm 0.062,$$

$$f_{Tu}^{(n)} = 0.014 \pm 0.003, \quad f_{Td}^{(n)} = 0.036 \pm 0.008, \quad f_{Ts}^{(n)} = 0.118 \pm 0.062,$$

and $f_{TG}^{(p,\,n)}$ is related to these values by

$$f_{TG}^{(p,\,n)} = 1 - \sum_{q=u,d,s} f_{Tq}^{(p,\,n)}. \tag{4.19}$$

Here, a_q is the effective coupling constant between the DM and the quark. We obtain the scattering cross-section (spin-independent) for the DM of a proton or neutron as

$$\sigma_{p,\,n}^{\text{SI}} = \frac{4m_r^2}{\pi} f_{p,\,n}^2, \tag{4.20}$$

where m_r is the reduced mass defined as $1/m_r = 1/m_{N_R^3} + 1/m_{p,\,n}$.

An approximate form of a_q/m_q can be recast in the following form:

$$\frac{a_q}{m_q} = \frac{y_{n_3}}{v\sqrt{2}} \left[\frac{1}{m_h^2} - \frac{1}{m_H^2} \right] \sin\alpha \cos\alpha, \tag{4.21}$$

where $y_{n_3} = \sqrt{2} m_{N_R^3}/v_{B-L}$ is the Yukawa coupling as specified in the second term of equation (4.11).

4.3 Scotogenic dark matter model

Another widely studied model for addressing DM and neutrino mass is the 'Scotogenic model'. The Greek word 'scotos' means 'darkness', hence the name 'Scotogenic'. In such models the neutrino mass is generated radiatively at one-loop from their coupling with DM. Scotogenic DM models are widely discussed in the literature [15–26] and have many variants.

Here, we would like to discuss the very basic type of Scotogenic model, where the SM is extended with one additional scalar doublet $\eta = (\eta^+, \eta^0)$ and three fermionic singlets N_i ($i = 1, 2, 3$). The model also postulates that the EW vacuum is invariant

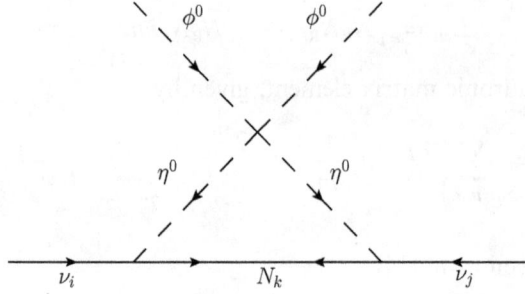

Figure 4.2. Radiative see-saw mechanism.

under a discrete Z_2-symmetry, under which all SM fields are even, whereas N_i and η are odd. As a result, η^0 has zero VEV and there is no Dirac mass linking ν_i with N_j. Neutrinos remain massless at tree level as in the SM.

The Yukawa interactions of this model are given by

$$\mathcal{L}_Y = f_{ij}(\phi^- \nu_i + \bar{\phi}^0 l_i)l_j^c + h_{ij}(\nu_i \eta^0 - l_i \eta^+)N_j + H.\,c. \tag{4.22}$$

In addition, there is a Majorana mass term of N_i written as $\frac{1}{2}M_i N_i N_i + H.\,c.$ The quartic scalar term $\frac{1}{2}\lambda_5(\Phi^\dagger \eta)^2 + H.\,c.$ is allowed. Hence the one-loop radiative generation of m_ν is possible, as shown in figure 4.2.

In this model, there are two possible DM candidates—either bosonic (the lightest of Re(η) or Im(η)) or fermionic (the lightest of $N_{1,2,3}$), whichever is the lightest stable (under Z_2-symmetry) particle. In either case the DM can be produced in the decay of the heavier one. For example, if the masses of N_i are heavier than that of η, we can observe the decay $N_i \to l^\pm \eta^\mp$ and subsequently $\eta^\mp \to \eta^0 W^\pm$.

Thus, the radiative see-saw model naturally predicts a DM candidate in a simple framework. These models have rich phenomenology and are testable in future collider experiments.

References

[1] Ma E 1998 Pathways to naturally small neutrino masses *Phys. Rev. Lett.* **81** 1171–4
[2] Kajiyma Y, Okada H and Toma T 2013 Light dark matter candidate in $B - L$ gauged radiative inverse seesaw *Eur. Phys. J.* C **73** 2381
[3] Okada N and Seto O 2010 Higgs portal dark matter in the minimal gauged $U(1)_{B-L}$ model *Phys. Rev.* D **82** 023507
[4] Kanemura S, Seto O and Shimomura T 2011 Masses of dark matter and neutrino from TeV scale spontaneous $U(1)_{B-L}$ breaking *Phys. Rev.* D **84** 016004
[5] Kanemura S, Nabeshima T and Sugiyama H 2012 TeV-scale seesaw with loop-induced Dirac mass term and dark matter from U(1)$_{B-L}$ gauge symmetry breaking *Phys. Rev.* D **85** 033004
[6] Okada H and Toma T 2012 Fermionic dark matter in radiative inverse seesaw model with $U(1)_{B-L}$ *Phys. Rev.* D **86** 033011
[7] Okada N and Orikasa Y 2012 Dark matter in the classically conformal $B - L$ model *Phys. Rev.* D **85** 115006

[8] Basak T and Mondal T 2014 Constraining minimal $U(1)_{B-L}$ model from dark matter observations *Phys. Rev.* D **89** 063527

[9] Kolb E W and Turner M S 1990 The early universe *Front. Phys.* **69** 1–547

[10] Beringer J *et al* 2012 Review of particle physics (RPP) *Phys. Rev.* D **86** 010001

[11] Srednicki M, Watkins R and Olive K A 1988 Calculations of relic densities in the early universe *Nucl. Phys.* B **310** 693

[12] Hinshaw G *et al* 2013 Nine-year Wilkinson microwave anisotropy probe (WMAP) observations: cosmological parameter results *Astrophys. J. Suppl.* **208** 19

[13] Ade P A R *et al* 2014 Planck 2013 results. XVI. Cosmological parameters *Astron. Astrophys.* **571** A16

[14] Ellis J R, Ferstl A and Olive K A 2000 Reevaluation of the elastic scattering of super-symmetric dark matter *Phys. Lett.* B **481** 304–14

[15] Ma E 2006 Verifiable radiative seesaw mechanism of neutrino mass and dark matter *Phys. Rev.* D **73** 077301

[16] Farzan Y and Ma E 2012 Dirac neutrino mass generation from dark matter *Phys. Rev.* D **86** 033007

[17] Ma E, Picek I and Radovi B 2013 New scotogenic model of neutrino mass with $U(1)_D$ gauge interaction *Phys. Lett.* B **726** 744–6

[18] Toma T and Vicente A 2014 Lepton flavor violation in the scotogenic model *J. High Energy Phys.* **01** 160

[19] Yu J-H 2016 Hidden gauged $U(1)$ model: unifying scotogenic neutrino and flavor dark matter *Phys. Rev.* D **93** 113007

[20] Ahriche A, McDonald K L and Nasri S 2016 The scale-invariant scotogenic model *J. High Energy Phys.* **06** 182

[21] Lindner M, Platscher M, Yaguna C E and Merle A 2016 Fermionic WIMPs and vacuum stability in the scotogenic model *Phys. Rev.* D **94** 115027

[22] Gu P-H, Ma E and Sarkar U 2016 Connecting radiative neutrino mass, neutron–antineutron oscillation, proton decay, and leptogenesis through dark matter *Phys. Rev.* D **94** 111701

[23] Wang W, Wang R, Han Z-L and Han J-Z 2017 The $B-L$ scotogenic models for Dirac neutrino masses *Eur. Phys. J.* C **77** 889

[24] Borah D and Gupta A 2017 New viable region of an inert Higgs doublet dark matter model with scotogenic extension *Phys. Rev.* D **96** 115012

[25] Tang Y-L 2018 Some phenomenologies of a simple scotogenic inverse seesaw model *Phys. Rev.* D **97** 035020

[26] Abada A and Toma T 2018 Electric dipole moments in the minimal scotogenic model *J. High Energy Phys.* **04** 030

Chapter 5

Supersymmetric dark matter model: a popular choice

The most popular and exhaustive extension of the Standard Model (SM) is supersymmetry (SUSY) (for reviews see [1–4]), which not only overcomes many shortcomings of the SM, but also has additional and very attractive features. This is a symmetry relating fermions to bosons such that for each fermionic degree of freedom there is a bosonic degree of freedom. This extends the particle content of the SM such that each particle in the SM has a corresponding superpartner (or partners). The major theoretical motivation behind SUSY is that it solves the hierarchy problem in a very simple and elegant way.

5.1 The hierarchy problem in the Standard Model

In the SM, the Higgs mass receives a contribution from the self-energy diagram, which turns out to be quadratically divergent. The scale of new physics, if it exists (i.e. the cut-off scale) is $\sim 10^{19}$ GeV, whereas the Higgs mass is at the electroweak scale ~ 100 GeV. Therefore, an unnatural fine-tuning [5, 6] is required to stabilize the weak scale against radiative corrections, if there is a high-scale physics relevant to the SM. This poses a serious problem for the SM, as this is an unnatural situation, which is caused by quadratic mass divergences in the scalar sector, known as the fine-tuning problem. In the context of grand unified theories (GUT), if we want to find a natural theory (i.e. not fine-tuned) in which scalar fields are associated with symmetry breaking, some symmetries are broken at the GUT scale $\sim 10^{16}$ GeV, whereas others are broken at the very much lower weak scale; this is usually referred to as the 'gauge hierarchy problem'.

This situation is dramatically improved by the introduction of SUSY. In supersymmetric theory, every divergent loop diagram containing an SM fermion is matched by corresponding scalar sfermion loop diagrams. Similarly, the contributions to the self-energy from SM particles and their superpartners would all cancel out, if SUSY were exact.

5.2 Minimal Supersymmetric Standard Model (MSSM)

The simplest supersymmetric extension of the SM is the MSSM [7–13], where the particle content is the same as that of the SM plus the superpartners and two Higgs doublets (instead of one as in the SM). Two Higgs doublets are needed for anomaly cancellation and to give mass to both up and down-type quarks and will result in five physical Higgs bosons. If SUSY were unbroken, an SM particle and its superpartner would have the same mass and quantum numbers (except for spin). Since we have not seen these particles, we can conclude that SUSY is broken at the energies probed by present accelerators.

The superpotential of the MSSM is given by

$$\mathcal{W} = \mu H_u \cdot H_d + y_u Q_{\mathrm{L}} \cdot H_u U_{\mathrm{R}} + y_d Q_{\mathrm{L}} \cdot H_d D_{\mathrm{R}} + y_e L_{\mathrm{L}} \cdot H_d E_{\mathrm{R}} \tag{5.1}$$

The scalar potential suggests that the bound on the lightest Higgs boson mass at tree level is $m_h \leqslant M_z |\cos 2\beta|$, which has been exceeded by LEP and LHC, where $m_h \geqslant 114.4$ GeV and 125–126 GeV, respectively. Therefore, a significant loop correction with maximal top-stop mixing is required to raise m_h up to 126 GeV.

5.2.1 Dark matter in MSSM

An additional discrete symmetry called R-parity is defined in the MSSM to evade proton-decay as $R = (1)^{2S+L+3B}$, with S, L and B the spin, leptonic and baryonic quantum number, respectively. We see that all SM particles have a parity of +1 while all supersymmetric particles have a parity of −1. If the R-parity is conserved throughout the interactions—i.e. if we impose this conservation by forcing a symmetry on the Lagrangian—then the lightest supersymmetric particle (LSP) will be stable and will not decay into lighter normal matter. This is a good way to ensure the stability at cosmological scales of DM. Hence, the LSP serves as a potential DM candidate.

The neutralinos are linear combinations of the superpartners of the neutral gauge bosons (i.e. gauginos) and the Higgs bosons (i.e. higgsinos). In the gauge-basis $(\tilde{B}, \tilde{W}^0, \tilde{H}_d^0, \tilde{H}_u^0)$, the mass matrix is given by

$$\mathcal{M}_{\tilde{G}} = \begin{pmatrix} M_1 & 0 & -\cos\beta \sin\theta_w M_Z & \sin\beta \sin\theta_w M_Z \\ 0 & M_2 & \cos\beta \cos\theta_w M_Z & -\sin\beta \cos\theta_w M_Z \\ -\cos\beta \sin\theta_w M_Z & \cos\beta \cos\theta_w M_Z & 0 & -\mu \\ \sin\beta \sin\theta_w M_Z & -\sin\beta \cos\theta_w M_Z & -\mu & 0 \end{pmatrix},$$

where M_1, M_2 are the soft breaking mass parameters for the bino and wino, respectively. Therefore, the neutralino mass matrix can be diagonalized analytically to give the four neutralinos

$$\tilde{\chi}_i^0 = N_{i1}\tilde{B} + N_{i2}\tilde{W}_3^0 + N_{i3}\tilde{H}_d^0 + N_{i4}\tilde{H}_u^0, \tag{5.2}$$

the lightest of which, $\tilde{\chi}_1^0$, serves as the main candidate for DM in SUSY models.

The lightest neutralino can be dominantly bino-like (in which case, $m_\chi \approx M_1$), wino-like (in which case, $m_\chi \approx M_2$), Higgsino-like (in which case, $m_\chi \approx \mu$) or a mixture of bino–Higgsino. If the LSP is dominantly higgsino-like it couples to the gauge boson very efficiently and thus will lead to a large pair annihilation cross-section. Hence, we could not achieve the desired relic abundance. If it is dominantly bino-like, which was the most favored scenario in SUSY models with gaugino mass unification at the GUT scale, one typically obtains too small an annihilation cross-section and, therefore, exceedingly large values of the DM relic density.

For a specific mass spectrum, where the LSP and the next heaviest super-symmetric particle (namely the next-to-lighter supersymmetric particle) are nearly degenerate, the co-annihilation process plays a crucial role in obtaining the relic abundance. Such processes include neutralino–chargino co-annihilation, which sets up the relic abundance for higgsino-like LSPs.

For wino-like LSPs, Sommerfeld enhancement [14, 15], which boosts the annihilation cross-section due to a modification of the Yukawa potential induced by the electroweak gauge bosons, plays a major role in obtaining correct relic density. This does not require any specific mass relation.

For bino-like LSPs, co-annihilation with stau (in the case $m_\chi \approx m_{\tilde{\tau}}$) [16, 17] can be crucial in determining the relic abundance. Similarly, the co-annihilation with the lightest squarks, such as lightest stop or sbottom, are important for heavy neutralinos.

Another very interesting phenomenon is the resonant enhancement of the annihilation cross-section via s-channel annihilation with Z, h and A as mediators, when the mediator mass is twice that of m_χ. This is known as the funnel-region. For, Z or h resonance the LSP is light (< 100 GeV), but can be much heavier in the case of pseudoscalar resonance.

5.2.2 Scalar interaction cross-section

The scalar interaction between the DM (i.e. neutralino LSP) and the quark is given by

$$\mathcal{L}_{\text{scalar}} = a_q \bar{\chi}\chi\bar{q}q, \tag{5.3}$$

where a_q is the coupling between the quark and the neutralino. The scalar cross-section for the neutralino scattering off a target nucleus (one has to sum over the proton and neutrons in the target) is given by

$$\sigma_{\text{scalar}} = \frac{4m_r^2}{\pi}(Zf_p + (A - Z)f_n)^2, \tag{5.4}$$

where m_r is the reduced mass of the nucleon and $f_{p,\,n}$ is the neutralino coupling to the proton or neutron [18, 19], given by

$$f_{p,n} = \sum_{q=u,d,s} f_{Tq}^{(p,n)} a_q \frac{m_{p,n}}{m_q} + \frac{2}{27} f_{TG}^{(p,n)} \sum_{q=c,b,t} a_q \frac{m_{p,n}}{m_q}, \qquad (5.5)$$

where $f_{Tu}^{(p)} = 0.020 \pm 0.004,$ $f_{Td}^{(p)} = 0.026 \pm 0.005,$ $f_{Ts}^{(p)} = 0.118 \pm 0.062,$ $f_{Tu}^{(n)} = 0.014 \pm 0.003,$ $f_{Td}^{(n)} = 0.036 \pm 0.008$ and $f_{Ts}^{(n)} = 0.118 \pm 0.062$ [20]. $f_{TG}^{(p,n)}$ is related to these values by

$$f_{TG}^{(p,n)} = 1 - \sum_{q=u,d,s} f_{Tq}^{(p,n)}. \qquad (5.6)$$

5.3 Next-to-Minimal Supersymmetric Standard Model (NMSSM)

Within the MSSM, loop corrections can give the required large corrections to the Higgs mass, provided the stop is heavier than 1 TeV or there is near maximal stop mixing. The implications of the 125 GeV Higgs for the MSSM and constrained MSSM parameter space have been extensively studied [7–13, 21–25]. Going beyond MSSM, in order to obtain a larger tree-level Higgs mass, the simplest extension is a singlet superfield in the NMSSM model [26–34]. The singlet interaction with the two Higgs doublet of MSSM is via the $\lambda S H_u \cdot H_d$ term.

The superpotential of NMSSM is given by

$$\mathcal{W} = \mu H_u \cdot H_d + y_u Q_L \cdot H_u U_R + y_d Q_L \cdot H_d D_R + y_e L_L \cdot H_d E_R + \lambda S H_u \cdot H_d + \kappa S^3. \quad (5.7)$$

The Higgs mass is now given by the relation $m_h^2 = M_Z^2 \cos^2 2\beta + \lambda^2 v^2 \sin^2 2\beta + \delta m_h^2$, where δm_h^2 is due to radiative correction. Taking $\lambda = 0.7$ (larger values would make it flow to the non-perturbative regime much below the GUT scale) and $\tan \beta = 2$ the radiative correction needed to obtain a 125 GeV Higgs mass is $\delta m_h = 55$ GeV which is an improvement over the $\delta m_h = 85$ GeV needed in the MSSM. However fine-tuning of the stop mass is still required in NMSSM to obtain the required Higgs mass [30, 35].

In NMSSM, the LSP can be a nearly pure singlino (the fermionic component of the singlet superfield S), which typically interacts very weakly with H_u and H_d. It would annihilate mainly into scalar–pseudoscalar pairs. The interaction strength would be proportional to the couplings κ or λ, which are typically small. Thus it would lead to an over-abundant scenario. Therefore, one needs to consider mixing of singlino–higgsino LSPs. However, this can be improved taking into consideration the co-annihilations with higgsino, wino, stau/sneutrino, stop, or a gluino [36].

5.4 Triplet extensions of the MSSM

The extension of the MSSM by extending it with $Y = 0$ and $Y = 0, \pm 1$ $SU(2)$ triplet superfields has been studied [37–39], where the tree-level contribution to the Higgs mass from the triplet Higgs sector has been estimated. It has been shown in [39] that with the $Y = 0$ triplet superfield the tree-level Higgs mass can be raised to 113 GeV, which would still require substantial loop corrections from stops. Recently, the

MSSM extended by two real triplets ($Y = \pm 1$) and one singlet [40] has been studied with a motivation to solve the μ-problem as well as to obtain a large correction to the lightest Higgs mass. The analysis of the DM sector of this model will be complicated as the LSP will be the lightest eigenstate of the 7×7 neutralino mass matrix, which has not yet been done.

5.4.1 Triplet–singlet MSSM

In this model [41], by taking naturalness of the Higgs mass as a guiding criterion, we have extended the superpotential of the MSSM by adding one singlet chiral superfield S and one $SU(2)$ triplet chiral superfield T_0 with hypercharge $Y = 0$. The most general form of the superpotential for this singlet–triplet extended model can be written as,

$$
W = (\mu + \lambda \hat{S})\hat{H}_d \cdot \hat{H}_u + \frac{\lambda_1}{3}\hat{S}^3 + \lambda_2\hat{H}_d \cdot \hat{T}_0\hat{H}_u + \lambda_3\hat{S}^2 \operatorname{Tr}(\hat{T}_0) + \lambda_4\hat{S}\operatorname{Tr}(\hat{T}_0\hat{T}_0) \tag{5.8}
$$
$$
+ W_{\text{Yuk}},
$$

where $\hat{H}_{u,d}$ are the Higgs doublets of the MSSM and the Yukawa superpotential W_{Yuk} is given as

$$
W_{\text{Yuk}} = y_u\hat{Q}_L \cdot \hat{H}_u\hat{U}_R + y_d\hat{Q}_L \cdot \hat{H}_d\hat{D}_R + y_e\hat{L}_L \cdot \hat{H}_d\hat{E}_R. \tag{5.9}
$$

We can solve the μ-problem by starting with a scale invariant superpotential, given as

$$
W_{\text{sc.inv.}} = \lambda\hat{S}\hat{H}_d \cdot \hat{H}_u + \frac{\lambda_1}{3}\hat{S}^3 + \lambda_2\hat{H}_d \cdot \hat{T}_0\hat{H}_u + \lambda_4\hat{S}\operatorname{Tr}(\hat{T}_0\hat{T}_0) + W_{\text{Yuk}}, \tag{5.10}
$$

where the $SU(2)$ invariant dot product is defined as

$$
\hat{H}_d \cdot \hat{T}_0\hat{H}_u = \frac{1}{\sqrt{2}}\left(\hat{H}_d^0\hat{T}^0\hat{H}_u^0 + \hat{H}_d^-\hat{T}^0\hat{H}_u^+\right) - \left(\hat{H}_d^0\hat{T}_0^-\hat{H}_u^+ + \hat{H}_d^-\hat{T}_0^+\hat{H}_u^0\right). \tag{5.11}
$$

By this choice we are eliminating the μ-parameter, but an effective μ-term is generated when the neutral components of S and T_0 acquire vacuum expectation values v_s and v_t, respectively,

$$
\mu_{\text{eff}} = \lambda v_s - \frac{\lambda_2}{\sqrt{2}}v_t. \tag{5.12}
$$

Neutralinos and charginos
In the fermionic sector, the neutral component of the triplet and singlet, i.e. $\tilde{T}^0$ and $\tilde{S}$, mix with the higgsinos and the gauginos. Thus, the neutralino mass matrix extended by the singlet and triplet sector, in the basis ($\tilde{B}$, $\tilde{W}^0$, $\tilde{H}_d^0$, $\tilde{H}_u^0$, $\tilde{S}$, $\tilde{T}^0$), is given by

$$\mathcal{M}_{\tilde{G}} = \begin{pmatrix} M_1 & 0 & -c_\beta s_w M_Z & s_\beta s_w M_Z & 0 & 0 \\ 0 & M_2 & c_\beta c_w M_Z & -s_\beta c_w M_Z & 0 & 0 \\ -c_\beta s_w M_Z & c_\beta c_w M_Z & 0 & -\mu_{\text{eff}} & -\lambda v_u & \dfrac{\lambda_2}{\sqrt{2}} v_u \\ s_\beta s_w M_Z & -s_\beta c_w M_Z & -\mu_{\text{eff}} & 0 & -\lambda v_d & \dfrac{\lambda_2}{\sqrt{2}} v_d \\ 0 & 0 & -\lambda v_u & -\lambda v_d & 2\lambda_1 v_s & 2\lambda_4 v_t \\ 0 & 0 & \dfrac{\lambda_2}{\sqrt{2}} v_u & \dfrac{\lambda_2}{\sqrt{2}} v_d & 2\lambda_4 v_t & 2\lambda_4 v_s \end{pmatrix}, \qquad (5.13)$$

where M_1, M_2 are the soft breaking mass parameters for the bino and wino, respectively, and

$$c_\beta = \cos \beta, \ s_\beta = \sin \beta, \ c_w = \cos \theta_w \ \text{and} \ s_w = \sin \theta_w.$$

The leftmost 4×4 entries are exactly identical with those in MSSM, except for the μ_{eff}-term which is defined in equation (5.12). As the triplet and the singlet fermion do not have any interaction with the neutral gauginos, the rightmost 2×2 entries are zero. The chargino mass terms in the Lagrangian can be written as

$$-\frac{1}{2}\left[\tilde{G}^{+T} M_c^T \cdot \tilde{G}^- + \tilde{G}^{-T} M_c \cdot \tilde{G}^+ \right], \qquad (5.14)$$

where the basis $\tilde{G}^+$ and $\tilde{G}^-$ are specified as

$$\tilde{G}^+ = \begin{pmatrix} \tilde{W}^+ \\ \tilde{H}_u^+ \\ \tilde{T}^+ \end{pmatrix}, \qquad \tilde{G}^- = \begin{pmatrix} \tilde{W}^- \\ \tilde{H}_d^- \\ \tilde{T}^- \end{pmatrix}.$$

Similarly, the charged component of the triplet $\tilde{T}^+$ and $\tilde{T}^-$ contribute to the chargino mass matrix. The chargino matrix in the gauge-basis $\tilde{G}^+$ and $\tilde{G}^-$ is given by

$$\mathcal{M}_{\text{ch}} = \begin{pmatrix} M_2 & \dfrac{1}{\sqrt{2}} g_2 v_d & g_2 v_t \\ \dfrac{1}{\sqrt{2}} g_2 v_u & \lambda v_s + \dfrac{\lambda_2}{\sqrt{2}} v_t & \lambda_2 v_d \\ -g_2 v_t & \lambda_2 v_u & 2\lambda_4 v_s \end{pmatrix}, \qquad (5.15)$$

where

$$\tilde{G}^+ = \begin{pmatrix} \tilde{W}^+ \\ \tilde{H}_u^+ \\ \tilde{T}^+ \end{pmatrix}, \qquad \tilde{G}^- = \begin{pmatrix} \tilde{W}^- \\ \tilde{H}_d^- \\ \tilde{T}^- \end{pmatrix}.$$

Since $\mathcal{M}_{ch}^T \neq \mathcal{M}_{ch}$, this matrix is diagonalised via bi-unitary transformation, which requires two distinct unitary matrices U and V such that

$$\tilde{\chi}^+ = V\tilde{G}^+,$$
$$\tilde{\chi}^- = U\tilde{G}^-.$$

(5.16)

The diagonal matrix reads

$$U^*\mathcal{M}_{ch}V^{-1} = \begin{pmatrix} m_{\tilde{\chi}_1^\pm} & 0 & 0 \\ 0 & m_{\tilde{\chi}_2^\pm} & 0 \\ 0 & 0 & m_{\tilde{\chi}_3^\pm} \end{pmatrix}$$

(5.17)

and similarly the Hermitian conjugate of equation (5.17) also gives the diagonal chargino mass matrix.

With the choice of a set of benchmark points, i.e. specifying the value of parameters at the electroweak scale, one can generate the mass spectrum of this model. For example, it may turn out that the LSP is a mixture of higgsino–triplino (triplet fermion) and serves as a viable DM candidate. Depending on the mass of DM, we need to analyze the possible annihilation channel, which may lead to gauge boson final states or fermionic final states. One can study the detailed phenomenology by analyzing the model parameter space with the help of the computational tools such as SuSpect [42], Spheno [43, 44], SARAH [45], DarkSUSY [46] and micrOmegas [47, 48].

References

[1] Martin S P 2010 A supersymmetry primer *Adv. Ser. Direct. High Energy Phys.* **21** 1–153
[2] Baer H and Tata X 2006 *Weak Scale Supersymmetry: from Superfields to Scattering Events* (Cambridge: Cambridge University Press)
[3] Aitchison I J R 2005 Supersymmetry and the MSSM: an elementary introduction, arXiv:hep-ph/0505105
[4] Haber H E and Kane G L 1985 The search for supersymmetry: probing physics beyond the standard model *Phys. Rep.* **117** 75–263
[5] Susskind L 1979 Dynamics of spontaneous symmetry breaking in the Weinberg–Salam theory *Phys. Rev.* D **20** 2619–25
[6] 't Hooft G, Itzykson C, Jaffe A, Lehmann H, Mitter P K, Singer I M and Stora R 1980 Recent developments in gauge theories *NATO Sci. Ser.* B **59** 1–438
[7] Baer H, Barger V and Mustafayev A 2012 Implications of a 125 GeV Higgs scalar for LHC SUSY and neutralino dark matter searches *Phys. Rev.* D **85** 075010
[8] Heinemeyer S, Stal O and Weiglein G 2012 Interpreting the LHC Higgs search results in the MSSM *Phys. Lett.* B **710** 201–6
[9] Arbey A, Battaglia M, Djouadi A, Mahmoudi F and Quevillon J 2012 Implications of a 125 GeV Higgs for supersymmetric models *Phys. Lett.* B **708** 162–9
[10] Draper P, Meade P, Reece M and Shih D 2012 Implications of a 125 GeV Higgs for the MSSM and low-scale SUSY breaking *Phys. Rev.* D **85** 095007

[11] Arbey A, Battaglia M and Mahmoudi F 2012 Constraints on the MSSM from the Higgs sector: a pMSSM study of Higgs searches, $B_s^0 \rightarrow \mu^+\mu^-$ and dark matter direct detection *Eur. Phys. J.* **C72** 1906

[12] Brummer F, Kraml S and Kulkarni S 2012 Anatomy of maximal stop mixing in the MSSM *J. High Energy Phys.* **1208** 089

[13] Nojiri M M, Polesello G and Tovey D R 2006 Constraining dark matter in the MSSM at the LHC *J. High Energy Phys.* **0603** 063

[14] Lattanzi M and Silk J I 2009 Can the WIMP annihilation boost factor be boosted by the Sommerfeld enhancement? *Phys. Rev.* D **79** 083523

[15] Hryczuk A, Iengo R and Ullio P 2011 Relic densities including Sommerfeld enhancements in the MSSM *J. High Energy Phys.* **03** 069

[16] Mizuta S and Yamaguchi M 1993 Coannihilation effects and relic abundance of Higgsino dominant LSP(s) *Phys. Lett.* B **298** 120–6

[17] Ellis J R, Falk T and Olive K A 1998 Neutralino–stau coannihilation and the cosmological upper limit on the mass of the lightest supersymmetric particle *Phys. Lett.* B **444** 367–72

[18] Jungman G, Kamionkowski M and Griest K 1996 Supersymmetric dark matter *Phys. Rep.* **267** 195–373

[19] Bertone G, Hooper D and Silk J 2005 Particle dark matter: evidence, candidates and constraints *Phys. Rep.* **405** 279–390

[20] Ellis J R, Ferstl A and Olive K A 2000 Reevaluation of the elastic scattering of super-symmetric dark matter *Phys. Lett.* B **481** 304–14

[21] Kadastik M, Kannike K, Racioppi A and Raidal M 2012 Implications of the 125 GeV Higgs boson for scalar dark matter and for the CMSSM phenomenology *J. High Energy Phys.* **1205** 061

[22] Cao J, Heng Z, Li D and Yang J M 2012 Current experimental constraints on the lightest Higgs boson mass in the constrained MSSM *Phys. Lett.* B **710** 665–70

[23] Ellis J and Olive K A 2012 Revisiting the Higgs mass and dark matter in the CMSSM *Eur. Phys. J.* **C72** 2005

[24] Baer H, Barger V and Mustafayev A 2012 Neutralino dark matter in mSUGRA/CMSSM with a 125 GeV light Higgs scalar *J. High Energy Phys.* **1205** 091

[25] Adeel Ajaib M, Gogoladze I, Nasir F and Shafi Q 2012 Revisiting mGMSB in light of a 125 GeV Higgs *Phys. Lett.* B **713** 462–8

[26] Drees M 1989 Supersymmetric models with extended Higgs sector *Int. J. Mod. Phys.* A **4** 3635

[27] Ellwanger U, Hugonie C and Teixeira A M 2010 The next-to-minimal supersymmetric standard model *Phys. Rep.* **496** 1–77

[28] Ross G G and Schmidt-Hoberg K 2012 The fine-tuning of the generalised NMSSM *Nucl. Phys.* B **862** 710–19

[29] Hall L J, Pinner D and Ruderman J T 2012 A natural SUSY Higgs near 126 GeV *J. High Energy Phys.* **1204** 131

[30] King S F, Muhlleitner M and Nevzorov R 2012 NMSSM Higgs benchmarks near 125 GeV *Nucl. Phys.* B **860** 207–44

[31] Kang Z, Li J and Li T 2012 On naturalness of the MSSM and NMSSM *J. High Energy Phys.* **1211** 024

[32] Cao J-J, Heng Z-X, Yang J M, Zhang Y-M and Zhu J-Y 2012 A SM-like Higgs near 125 GeV in low energy SUSY: a comparative study for MSSM and NMSSM *J. High Energy Phys.* **1203** 086

[33] Vasquez D A *et al* 2012 The 125 GeV Higgs in the NMSSM in light of LHC results and astrophysics constraints *Phys. Rev.* D **86** 035023

[34] Ellwanger U and Hugonie C 2012 Higgs bosons near 125 GeV in the NMSSM with constraints at the GUT scale *Adv. High Energy Phys.* **2012** 625389

[35] Cao J-J *et al* 2011 Light dark matter in NMSSM and implication on Higgs phenomenology *Phys. Lett.* B **703** 292–7

[36] Belanger G, Boudjema F, Hugonie C, Pukhov A and Semenov A 2005 Relic density of dark matter in the NMSSM *J. Cosmol. Astropart. Phys.* **0509** 001

[37] Espinosa J R and Quiros M 1992 On Higgs boson masses in nonminimal supersymmetric standard models *Phys. Lett.* B **279** 92–7

[38] Espinosa J R and Quiros M 1992 Higgs triplets in the supersymmetric standard model *Nucl. Phys.* B **384** 113–46

[39] Di Chiara S and Hsieh K 2008 Triplet extended supersymmetric standard model *Phys. Rev.* D **78** 055016

[40] Agashe K, Azatov A, Katz A and Kim D 2011 Improving the tunings of the MSSM by adding triplets and singlet *Phys. Rev.* D **84** 115024

[41] Basak T and Mohanty S 2012 Triplet–singlet extension of the MSSM with a 125 Gev Higgs and dark matter *Phys. Rev.* D **86** 075031

[42] Djouadi A, Kneur J-L and Moultaka G 2007 SuSpect: a Fortran code for the supersymmetric and Higgs particle spectrum in the MSSM *Comput. Phys. Commun.* **176** 426–55

[43] Porod W 2003 SPheno, a program for calculating supersymmetric spectra, SUSY particle decays and SUSY particle production at $e^+ e^-$ colliders *Comput. Phys. Commun.* **153** 275–315

[44] Porod W and Staub F 2012 SPheno 3.1: extensions including flavour, CP-phases and models beyond the MSSM *Comput. Phys. Commun.* **183** 2458–69

[45] Staub F 2014 SARAH 4: a tool for (not only SUSY) model builders *Comput. Phys. Commun.* **185** 1773–90

[46] Gondolo P, Edsjo J, Ullio P, Bergstrom L, Schelke M and Baltz E A 2004 DarkSUSY: computing supersymmetric dark matter properties numerically *J. Cosmol. Astropart. Phys.* **0407** 008

[47] Belanger G, Boudjema F, Pukhov A and Semenov A 2014 micrOMEGAs 3: a program for calculating dark matter observables *Comput. Phys. Commun.* **185** 960–85

[48] Belanger G, Boudjema F, Pukhov A and Semenov A 2010 micrOMEGAs: a tool for dark matter studies *Nuovo Cim.* **C033N2** 111–6

Chapter 6

FIMP as dark matter

Over the past decade, WIMP dark matter (DM) models have gained a lot of popularity due to their simplicity and economic framework. But, the lack of evidence for WIMP DM has prompted interest in other DM candidates and their identification, which usually requires search strategies very different to those for WIMPs. In this chapter, we discuss an alternative model to the popular WIMP model, known as the feebly interacting massive particle (FIMP).

6.1 Feebly interacting massive particle

Due to the non-observance of WIMP DM candidates in the direct detection experiments, a considerable amount of parameter space has already been excluded. In this scenario, the FIMP [1–8] appears as a viable alternative to the WIMP DM candidate. The observed abundance of DM is generated through the out-of-equilibrium process, by the so-called freeze-in mechanism.

6.1.1 Freeze-in mechanism

In the case of FIMP, the interaction with the visible bath particles is extremely weak. Hence, it is very weakly coupled to the thermal bath. The coupling strength is of the order of 10^{-7} or less. Hence, it never attains chemical equilibrium with the Standard Model (SM) particles in the thermal soup. Therefore, the abundance of FIMPs is never set-up by the well-known thermal freeze-out process, as in the case of WIMPs. Rather, it is either produced by the decay or $2 \to 2$ annihilation process of visible bath particles. The initial abundance of FIMP DM is assumed to be negligible. If $2m_\chi < m_\phi$, the FIMP abundance is achieved by the decay of the visible bath particle into a DM particle, i.e. $\phi \to \chi\chi$. This process continues until the number density of ϕ is Boltzmann suppressed. The comoving number density of DM particles then becomes a constant and the DM abundance freezes in (figure 6.1).

doi:10.1088/978-1-64327-132-3ch6   6-1

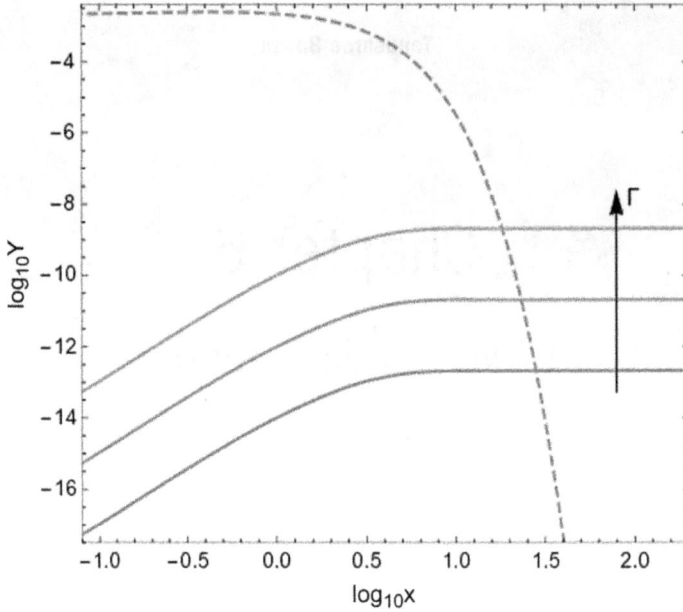

Figure 6.1. The two basic mechanisms for DM production: the freeze-out (left panel) and freeze-in (right panel), for three different values of the interaction rate between the visible sector and DM particles. Reproduced with permission from [8]. Copyright 2017 World Scientific.

The Boltzmann equation is given by [1]

$$\frac{x}{Y_\phi^{eq}}\frac{dY}{dx} = 2\frac{K_1(x)}{K_2(x)}\frac{\Gamma_{\phi\to\chi\chi}}{H}, \tag{6.1}$$

where K_i are the modified Bessel function, $Y = n_\chi/s$ and $x = m_\phi/T$. The decay width of ϕ is denoted by $\Gamma_{\phi\to\chi\chi}$. Solving the above equation, the relic abundance can be expressed as

$$\Omega_\chi h^2 \simeq 4.48 \times 10^8 \frac{g_\phi}{g_\star^{3/2}}\left(\frac{m_\chi}{\text{GeV}}\right)\frac{M_P\Gamma_{\phi\to\chi\chi}}{m_\phi^2}, \tag{6.2}$$

where g_ϕ is the internal degree of freedom of the field ϕ.

6.1.2 Higgs portal FIMP model

Let us consider a Higgs portal FIMP model, with DM as singlet scalar χ. Therefore, the interaction term between the SM-Higgs and the DM is given by

$$L_{int} = \frac{y_h}{2}|\phi|^2\chi^2, \tag{6.3}$$

where y_h is the coupling strength between the DM and Higgs boson. In the case of fermion or gauge boson DM, there would have been non-renormalisable interaction terms.

To ensure stability of the DM candidate, additional Z_2 symmetry has been imposed. All the SM particles are even under Z_2, except χ. The most general scalar potential can be written as

$$V(\phi, \chi) = V_{SM} + \frac{\mu^2}{2}\chi^2 + \frac{y_h}{2}|\phi|^2\chi^2 + \frac{y_\chi}{4}\chi^4, \tag{6.4}$$

where y_χ is the FIMP self-interaction coupling. If, $2m_\chi < m_h$, the DM is produced from the decay process $h \rightarrow \chi\chi$. Roughly, the decay width can be estimated as $\Gamma_{\phi \rightarrow \chi\chi} \simeq y_h^2 m_h/(8\pi)$. Therefore, the coupling y_h can be constrained in order to obtain correct relic abundance

$$\frac{\Omega_\chi h^2}{0.12} \simeq 10^{21} y_h^2 \left(\frac{m_\chi}{GeV}\right). \tag{6.5}$$

Hence, the bound on the coupling is $y_h \lesssim 10^{9-10}$. In the case of ϕ being any generic scalar particle other than the SM-Higgs, the parameter space of this model can studied for different values of m_ϕ and y_h.

6.1.3 Detection prospects

Due to the feeble coupling it seems challenging to detect the FIMP candidates in direct detection experiments. However, the low threshold direct detection experiments can probe such DM candidates in the mass range of keV to MeV [2, 9, 10].

Similarly, by virtue of their low interaction cross-section, such DM candidates are long-lived. Therefore, it is possible to address the decaying DM within this framework. Interpretation of the spectral feature at $E \sim 3.55$ keV observed in x-ray observations [11] from several DM dominated sources has been widely studied in the context of frozen-in decaying DM [12–14].

References

[1] Hall L J, Jedamzik K, March-Russell J and West S M 2010 Freeze-in production of FIMP dark matter *J. High Energy Phys.* **03** 080

[2] Essig R, Mardon J and Volansky T 2012 Direct detection of sub-GeV dark matter *Phys. Rev. D* **85** 076007

[3] Yaguna C E 2011 The singlet scalar as FIMP dark matter *J. High Energy Phys.* **08** 060

[4] Biswas A, Choubey S and Khan S 2017 Neutrino mass, leptogenesis and FIMP dark matter in a U(1)$_{B-L}$ model *Eur. Phys. J. C* **77** 875

[5] Banik A D, Pandey M, Majumdar D and Biswas A 2017 Two component WIMP-FImP dark matter model with singlet fermion, scalar and pseudo scalar *Eur. Phys. J. C* **77** 657

[6] Hessler A G, Ibarra A, Molinaro E and Vogl S 2017 Probing the scotogenic FIMP at the LHC *J. High Energy Phys.* **01** 100

[7] Molinaro E, Yaguna C E and Zapata O 2014 FIMP realization of the scotogenic model *J. Cosmol. Astropart. Phys.* **1407** 015

[8] Bernal N, Heikinheimo M, Tenkanen T, Tuominen K and Vaskonen V 2017 The dawn of FIMP dark matter: a review of models and constraints *Int. J. Mod. Phys. A* **32** 1730023

[9] Essig R, Manalaysay A, Mardon J, Sorensen P and Volansky T 2012 First direct detection limits on sub-GeV dark matter from XENON10 *Phys. Rev. Lett.* **109** 021301

[10] Essig R, Fernandez-Serra M, Mardon J, Soto A, Volansky T and Yu T T 2016 Direct detection of sub-GeV dark matter with semiconductor targets *J. High Energy Phys.* **5** 46

[11] Boyarsky A, Ruchayskiy O, Iakubovskyi D and Franse J 2014 Unidentified line in x-ray spectra of the Andromeda galaxy and Perseus galaxy cluster *Phys. Rev. Lett.* **113** 251301

[12] Roland S B, Shakya B and Wells J D 2015 PeV neutrinos and a 3.5 keV x-ray line from a PeV-scale supersymmetric neutrino sector *Phys. Rev.* D **92** 095018

[13] Bomark N E and Roszkowski L 2014 The 3.5 keV x-ray line from decaying gravitino dark matter *Phys. Rev.* D **90** 011701

[14] Merle A and Schneider A 2015 Production of sterile neutrino dark matter and the 3.5 keV line *Phys. Lett.* B **749** 283–8

Chapter 7

Conclusion and outlook

The identity of dark matter (DM) is one of the key outstanding problems in both particle physics and astrophysics. The presence of DM has been supported by a variety of evidence. At galactic and sub-galactic scales, this evidence includes galactic rotation curves, the weak gravitational lensing of distant galaxies by foreground structure, and the weak modulation of strong lensing around individual massive elliptical galaxies. On cosmological scales, observations of the anisotropies in the cosmic microwave background and large scale structure strongly lead us to the conclusion that 80–85% of the matter in the Universe (by mass) consists of non-luminous and non-baryonic matter. Here arises the pressing need to address the fundamental question about the properties and nature of DM. One needs to look beyond the Standard Model in order to accommodate a suitable DM candidate. There are several experimental constraints on DM, which include relic density measurements from WMAP and PLANCK, and direct detection and indirect detection experiments. The mass and scattering cross-section of the DM of the nucleon is probed by direct detection experiments such as XENON100, CDMS, DAMA, CoGENT, LUX, etc. The indirect detection experiments, such as PAMELA, AMS02 and Fermi-LAT, rely on the observations of DM pair annihilation into positrons, antiprotons and photons, which might indicate the existence of DM.

This book has provided a pedagogical overview of properties and experimental constraints on DM. It has also introduced the variety of the particle DM candidate zoo. Gradually, this book has taken you on a journey through the pathways of different DM models. This book primarily aims to teach how to build a successful particle physics model of DM, starting from the easiest extension of the SM with a gauge singlet to two-component DM scenario. This book has also focused on models addressing the fundamental issue of both DM and neutrino mass in unified scenario. Finally we have covered the widely studied supersymmetric DM models starting from the simplest MSSM to the singlet and triplet extension models. Chapters 3, 4 and 5 focussed on the WIMP DM candidate. In each of the models,

doi:10.1088/978-1-64327-132-3ch7

the aim is to accommodate a viable DM candidate and study its phenomenological aspects. As a detour to the WIMP model, an alternative choice of feebly interacting massive particle (FIMP) models was discussed in chapter 6. The production mechanism of FIMPs is very different to that of WIMPs, and has grabbed lot of attention recently. Although there are plenty of non-WIMP models (such as axion DM, sterile neutrino DM, gravitino DM, etc) in the literature, from the model-building perspective WIMP models are preferred. Hence, a major part of this book has focussed on model-building aspects of WIMPs.

DM itself is a very vast field of research from both the theoretical and experimental point of view. Innumerable models of DM have been proposed in the literature so far. This book, for the first time, has tried to provide a glimpse of a few different kinds of model-building perspectives. Having completed these six chapters, on different flavors of DM models, one should be able to grasp the basic aspects of model building.

It is a rather challenging task to build a successful DM model while complying with all the observational constraints at the same time. This will remain an exciting field of research as the future holds many more surprises in the form of experimental evidence and data. Therefore, the need for proposing or building new models of DM will never be exhausted.

Appendix A

A.1 Calculation of $w(s)$

Let ϕ be the scattering angle between incoming DM particles. Then $w(s)$ can be defined as

$$w(s) = \frac{1}{32\pi} \sqrt{\frac{s - 4m_{\text{final}}^2}{s}} \int \frac{d \cos \phi}{2} \sum_{\text{all possible channels}} |M|^2. \tag{A.1}$$

The function $|\mathcal{M}|^2$ contains not only the interaction part, but also contains the kinematical part. Considering the processes as in equation (4.16) we can write

$$
\begin{aligned}
w(s)_{b,\,\tau,\,W,\,Z} = &\left[\frac{\sin^2 \alpha \cos^2 \alpha}{4} \left(4y_{n_3}^2 (s - 4m_{N_R^3}^2) \right) \right] \\
&\times \left[\frac{1}{(s - m_h^2)^2 + \Gamma_h^2 m_h^2} + \frac{1}{(s - m_H^2)^2 + \Gamma_H^2 m_H^2} \right. \\
&\left. - 2 \frac{(s - m_h^2)(s - m_H^2) + m_h m_H \Gamma_h \Gamma_H}{((s - m_h^2)^2 + \Gamma_h^2 m_h^2)((s - m_H^2)^2 + \Gamma_H^2 m_H^2)} \right] \\
&\times \left[\left\{ \frac{1}{8\pi} \sqrt{\frac{s - m_b^2}{s}} \, 4y_b^2 \left(\frac{s}{4} - m_b^2 \right) 3 \right\} \right. \\
&+ \left\{ \frac{1}{8\pi} \sqrt{\frac{s - m_\tau^2}{s}} \, 4y_\tau^2 \left(\frac{s}{4} - m_\tau^2 \right) \right\} \\
&+ \left\{ \frac{1}{8\pi} \sqrt{\frac{s - m_W^2}{s}} \left(\frac{2m_W^2}{v} \left(s + \frac{1}{2m_W^4} \left(\frac{s}{2} - m_W^2 \right) \right) \right) \right\} \\
&+ \left. \left\{ \frac{1}{8\pi} \sqrt{\frac{s - m_Z^2}{s}} \left(\frac{m_Z^2}{v} \left(s + \frac{1}{2m_Z^4} \left(\frac{s}{2} - m_Z^2 \right) \right) \right) \right\} \right].
\end{aligned}
\tag{A.2}
$$

In this expression, the second line is the propagator function which includes both h and H. The third line shows the decay cross-section to $b\bar{b}$ and $\tau^+ \tau^-$, whereas the fourth and fifth lines are the decay cross-sections to $W^+ W^-$ and ZZ, respectively. In addition, we have also considered the annihilation into the SM-like Higgs bosons, for which $w(s)_h$ is given by

$$
\begin{aligned}
w(s)_h = &\left\{ \frac{1}{16\pi} \left[4y_{n_3}^2 (s - 4m_{N_R^3}^2) \right] \sqrt{\frac{s - m_h^2}{s}} \right. \\
&\left(\left(\frac{\sin\alpha}{\sqrt{2}} \right)^2 \frac{\lambda_{hhh}^2}{(s - m_h^2)^2 + \Gamma_h^2 m_h^2} + \left(\frac{\cos\alpha}{\sqrt{2}} \right)^2 \frac{\lambda_{Hhh}^2}{(s - m_H^2)^2 + \Gamma_H^2 m_H^2} \right. \\
&\left. \left. - \frac{\sin\alpha \ \cos\alpha \ \lambda_{hhh} \ \lambda_{Hhh} \ \{(s - m_h^2)(s - m_H^2) + m_h m_H \Gamma_h \Gamma_H\}}{((s - m_h^2)^2 + \Gamma_h^2 m_h^2) \ ((s - m_H^2)^2 + \Gamma_H^2 m_H^2)} \right) \right\},
\end{aligned}
\tag{A.3}
$$

where λ_{hhh} and λ_{Hhh} are calculated by expanding the Higgs potential part,

$$
\begin{aligned}
\lambda_{hhH} = &\ 3\lambda_1 v (\cos^2\alpha \sin\alpha) + 3\lambda_2 v_{B-L} (\cos\alpha \sin^2\alpha) \\
&+ \frac{1}{8}\lambda_3 \{ v_{B-L}(\cos\alpha + 3\cos(3\alpha)) + v(\sin\alpha - 3\sin(3\alpha)) \}, \\
\lambda_{hhh} = &\ \frac{\lambda_1}{4} v (3\cos\alpha + \cos(3\alpha)) + \frac{\lambda_2}{4} v_{B-L} (-3\sin\alpha + \sin(3\alpha)) \\
&+ \frac{\lambda_3}{8}(v(\cos\alpha - \cos(3\alpha)) - v_{B-L}(\sin\alpha + \sin(3\alpha))).
\end{aligned}
\tag{A.4}
$$

Finally, $w(s) = w(s)_{b, \ \tau, \ W, \ Z} + w(s)_h$.